The geographer's vademecum

of sources and materials

compiled by

J. C. HANCOCK, B.Sc., and P. F. WHITELEY, B.A.

LONDON: GEORGE PHILIP AND SON LIMITED

CONTENTS

J. C. Hancock, B.Sc., formerly Senior Lecturer in Geography, St. Paul's College, Cheltenham, is now Research Associate, School of Education, University of Bristol.
P. F. Whiteley, B.A., is Senior Lecturer in Geography, St. Mary's College, Cheltenham.

Every care has been exercised in the preparation of this publication from information received, but neither the compilers nor the publishers can accept responsibility for inaccuracies therein.

Printed in Great Britain by
William Clowes and Sons Limited,
London, Colchester and Beccles

SBN 540 00209 7

PREFACE

The task of the modern teacher is complicated by the tremendous increase in the amount of detail and factual information about his subject, and by changing methods of teaching which often involve children in more activity, whether through 'discovery' work, environmental studies or projects.

This book seeks to meet the need for a convenient ready-reference key to some of the many sources of information and material, much of it from organisations outside the field of educational publishing and supply as such, that are available to teachers of geography and related subjects. To attempt a comprehensive coverage would be impracticable, but the compilers' aim has been to include as much material as possible of a kind that meets the twin criteria of relevance and usefulness. Every organisation listed has been contacted, and all have expressed their willingness to help active teachers and teachers in training.

Part One of the book, compiled by P. F. Whiteley, is divided into two sections. Section 'A' is an alphabetical list of selected *topics* with the names of relevant organisations under each. Section 'B' lists these organisations and their addresses and gives details of materials available from each one. Part Two, compiled by J. C. Hancock, is arranged alphabetically by *regions and countries* and gives particulars of selected sources and materials under each. At the end of the book there is a key to organisations whose names, because of their frequent occurrence, are abbreviated or appear without addresses in the lists.

The compilers would like to thank all the firms and organisations who have responded so helpfully to their enquiries, and they gratefully acknowledge the invaluable help and encouragement of the Principals of St. Paul's and St. Mary's Colleges, Cheltenham, and several of the students of these colleges; Mr. E. R. Burrell; Mr. J. C. Milner; and Dr. G. Hickman of the Geography Panel of the School of Education, University of Bristol.

J. C. H. P. F. W.

Writing for information and material

Many of the sources listed in this book are not educational publishers or suppliers. In order to make the best use of the services offered by such organisations, letters should request assistance on specific topics. In some cases material is only supplied to Head Teachers; these entries are preceded by the symbol †. A few organisations, indicated by an asterisk, are prepared to accept enquiries from children, but for obvious reasons repetitive, mass enquiries should always be avoided.

Films and filmstrips

Unless otherwise stated, films are 16 mm, in colour and with a sound commentary; filmstrips are 35 mm, in colour, and single-frame.

Maps

Unless otherwise stated, the prices given for maps listed under *Wallmaps and charts* refer to cloth-mounted versions; in some cases they are of course also available on paper at less cost.

Prices

Prices are continuously changing and it is important to note that they are included in this book only as a broad guide to give the user some idea of cost; current prices should always be confirmed before ordering. In many cases price changes obtaining this year are not available at the time of going to press, and to give decimal conversions of 1970 prices could therefore be very misleading. In view of this prices are quoted in £ s. d. throughout, but for convenience an index of £p equivalents of all the prices in the book is given below.

s.	d.	£p	s.	d.	£p	s.	d.	£p	s.	d.	£p
—	6	2½	7	6	37½	18	—	90	55	—	2.75
—	8	3½	8	—	40	21	—	1.05	55	3	2.76
—	9	4	8	4	41½	22	—	1.10	57	6	2.87½
1	—	5	8	6	42½	22	6	1.12½	62	6	3.12½
1	1	5½	9	—	45	23	—	1.15	63	—	3.15
1	2	6	10	—	50	24	—	1.20	63	9	3.19
1	3	6	10	6	52½	25	—	1.25	67	6	3.37½
1	6	7½	10	11	54½	26	—	1.30	69	—	3.45
1	9	9	11	—	55	26	6	1.32½	70	—	3.50
1	10	9	11	6	57½	28	6	1.42½	71	3	3.56
2	—	10	11	7	58	29	9	1.49	78	—	3.90
2	3	11	11	10	59	30	—	1.50	85	—	4.25
2	6	12½	12	—	60	34	—	1.70	95	—	4.75
2	9	14	12	6	62½	36	—	1.80	97	6	4.87½
2	10	14	13	—	65	37	6	1.87½	105	—	5.25
3	—	15	13	6	67½	39	—	1.95	110	—	5.50
3	6	17½	14	—	70	39	6	1.97½	112	6	5.62½
3	9	19	14	6	72½	41	—	2.05	130	—	6.50
4	—	20	15	—	75	42	—	2.10	170	—	8.50
5	—	25	15	6	77½	44	—	2.20	175	—	8.75
5	6	27½	15	7	78	50	—	2.50	185	—	9.25
6	—	30	16	—	80	51	—	2.55	210	—	10.50
6	6	32½	16	6	82½	52	3	2.61			
7	—	35	17	6	87½	52	6	2.62½			

Based on the *Shoppers' Table* issued by the Decimal Currency Board

Part One

[A] TOPICS AND SOURCES

Agriculture
Association of Agriculture
Danish Agricultural Producers Information Service
Department of Agriculture and Fisheries for Scotland
European Community Information Service
Farmer and Stockbreeder Publications Ltd.
Ministry of Agriculture, Fisheries and Food
National Farmers Union
Platignum Pen Co. Ltd.
Scottish Association of Young Farmers Clubs

Airlines and air transport
British Aircraft Corporation Ltd.
British Overseas Airways Corporation
Educational Productions Ltd.
Pan Am Publications

Aluminium
Aluminium Federation
British Aluminium Co. Ltd.

Asbestos
Commonwealth Institute

Atomic power
European Community Information Service
United Kingdom Atomic Energy Authority

Bananas
Elders and Fyffes Ltd.

Bauxite *see* Aluminium

Beverages *see* Tea, Coffee, Cocoa

Biscuits
Associated Biscuits Ltd.

Building materials
Cement and Concrete Association
Cement Marketing Co. Ltd.
National Federation of Clay Industries

Butter

Butter Information Council
Danish Agricultural Producers Information Service
Dutch Dairy Bureau
National Dairy Council
New Zealand Dairy Board

Canals

Inland Waterways Association Ltd.
Manchester Ship Canal Co.

Cement

Cement and Concrete Association
Cement Marketing Co. Ltd.

Cereals (*see also* Wheat, Rice, Maize)

Associated Biscuits Ltd.
Kellogg Company of Great Britain Ltd.

Cheese

Cheese Bureau
Dutch Dairy Bureau
Kraft Foods Ltd.
National Dairy Council
New Zealand Dairy Board

Chemicals

Imperial Chemical Industries Ltd.

Chocolate

J. S. Fry and Sons Ltd.
Nestlé Co. Ltd.
Rowntree and Co. Ltd.

Coal

National Coal Board

Cocoa

Commonwealth Institute
Rowntree and Co. Ltd.

Coffee

Coffee Promotion Council Ltd.
Commonwealth Institute
J. Lyons and Co. Ltd.
Nestlé Co. Ltd.

Common Market

European Community Information Service

Concrete

Cement and Concrete Association

Copper

Commonwealth Institute
Copper Development Association
Roan Selection Trust Group

Cotton

Commonwealth Institute
Textile Council

Dairy produce and dairy farming

Association of Agriculture
Butter Information Council
Danish Agricultural Producers Information Service
Dutch Dairy Bureau
Milk Marketing Board
National Dairy Council
New Zealand Dairy Board

Diamonds

De Beers Consolidated Mines Ltd.
Industrial Diamond Information Bureau

Electricity

Central Electricity Generating Board
Electricity Council
Joseph Lucas Ltd.

Engineering

British Leyland Motor Corporation Ltd.
Joseph Lucas Ltd.
Shipbuilders and Repairers National Association
Smiths Industries Ltd.
Volkswagen Motors Ltd.

Farming *see* Agriculture, Cheese, Dairy Produce, Milk, Fruit

Fire

Joseph Lucas Ltd.

Fish

Fish Education Service of the White Fish Authority
Herring Industry Board
Unilever Ltd.

Flax *see* Linen

Folk music

Topic Records Ltd.

Food canning

Heinz Home Cookery Service
Smedleys Ltd.

Fruit

Association of Agriculture
Commonwealth Institute
Deciduous Fruit Board
Elders and Fyffes Ltd.
Jaffa Information Bureau
Libby McNeill and Libby Ltd.
National Federation of Fruit and Potato Traders Ltd.
Outspan Organisation
James Robertson and Sons Ltd.
Southern Postal and Duplicating Services

Furniture

Dunlop Rubber Co. Ltd.

Furs

Hudsons Bay Co.

Geology

Shell International Petroleum Co. Ltd.

Glass

Glass Manufacturers Federation
Pilkington Bros. Ltd.

Gypsum

British Gypsum Ltd.
Gypsum Products Development Association

Homes

Cosmic Crayon Co. Ltd.
Building Societies Association

Insurance

British Insurance Association

Iron and steel

British Steel Corporation

Jute

Association of Jute Spinners and Manufacturers
Commonwealth Institute

Kaolin (*see also* Pottery)

English China Clays Group

Lead

Lead Development Association
A. Strauss and Co. Ltd.

Leather

Leather Institute
Norvic Shoe Co. Ltd.

Lifeboats

Royal National Lifeboat Institution

Light

Joseph Lucas Ltd.

Lighthouses and lightships

Trinity House Lighthouse Service

Limestone

Cement and Concrete Association
Cement Marketing Co. Ltd.
I.C.I. Ltd.

Linen

Flaxspinners and Manufacturers Association of Great Britain
Irish Linen Guild

Local geography

Cosmic Crayon Co. Ltd.
Platignum Pen Co. Ltd.

Maize

Commonwealth Institute

Man-made fibres

British Enkalon Ltd.
British Man-made Fibres Federation
Courtaulds Ltd.
Du Pont Co. (U.K.) Ltd.
I.C.I. Fibres Ltd.
Monsanto Textiles Ltd.

Margarine

Unilever Ltd.
Van Den Berghs Ltd.

Milk

Association of Agriculture
Milk Marketing Board
National Dairy Council

Motor vehicles

British Leyland Motor Corporation
Ford Motor Co. Ltd.
Volkswagen Motors Ltd.

National Parks

Countryside Commission

Natural gas

Gas Council Film Library
Petroleum Information Bureau
Shell International Petroleum Co. Ltd.

Oil *see* Petroleum

Olympic Games

E. J. Arnold and Son Ltd.

Oranges

Jaffa Information Bureau
Outspan Organisation

Palm oil

Commonwealth Institute
Unilever Ltd.

Paper (*see also* Timber)

Commonwealth Institute
Reed Paper Group Ltd.
Wiggins Teape and Co. Ltd.

Peaches

Libby McNeill and Libby Ltd.

Petroleum

British Petroleum Co. Ltd.
Esso Petroleum Co. Ltd.
Petroleum Information Bureau
Shell International Petroleum Co. Ltd.

Plastics

I.C.I. Ltd. (Plastics Division)

Population

Catholic Institute for International Relations
Christian Aid
Foreign & Commonwealth Office
(Overseas Development Administration)
Overseas Development Institute Ltd.

continued

Oxfam
United Nations Information Centre
Voluntary Committee on Overseas Aid and Development

Ports

British Ship Adoption Society
British Transport Docks Board, Hull
British Transport Docks Board, Southampton
Manchester Ship Canal Co.
Mersey Docks and Harbour Board
Port of Bristol Authority
Port of London Authority
Scouse Press

Potatoes

National Federation of Fruit and Potato Trades Ltd.
Potato Marketing Board

Pottery

British Pottery Promotion Service Ltd.
Doulton and Co. Ltd.
English China Clays Group
Josiah Wedgwood and Sons Ltd.

Projects

E. J. Arnold and Son Ltd.
British Trades Alphabet
Cosmic Crayon Co. Ltd.
Hamish Hamilton Ltd.
I-Spy
Platignum Pen Co. Ltd.

Railways

British Railways Board
Canadian Pacific

Rice

Commonwealth Institute
Kellogg Company of Great Britain Ltd.
Rice Council

Rubber

Commonwealth Institute
Dunlop Rubber Co. Ltd.
Goodyear Tyre and Rubber Co. (Great Britain) Ltd.
Natural Rubber Producers Research Association

Salt

I.C.I. Ltd.

Ships

E. J. Arnold and Son Ltd.
British Ship Adoption Society
Cosmic Crayon Co. Ltd.
Dunlop Rubber Co. Ltd.
Furness Withy and Co. Ltd.
New Zealand Shipping Co. Ltd.
Royal National Lifeboat Institution
Shipbuilders and Repairers National Association
Union-Castle Mail Steamship Co. Ltd.

Silk

Lullingstone Silk Farm Ltd.
Silk Educational Service

Sisal

Commonwealth Institute

Soap

Unilever Ltd.

Spices

Commonwealth Institute

Steel *see* Iron and steel

Sugar

British Sugar Bureau
Commonwealth Institute

Tea

Brooke Bond Tea Ltd.
Ceylon Tea Centre
Commonwealth Institute
J. Lyons and Co. Ltd.

Textiles *see* Wool, Cotton, Jute, Linen, Man-made fibres

Timber

Commonwealth Institute
Forestry Commission
Timber Research and Development Association
Wiggins Teape and Co. Ltd.

Tin

Commonwealth Institute

Tobacco

Commonwealth Institute
J. Player and Sons
W. D. & H. O. Wills

Town planning
Town and Country Planning Association

Towns (*see also* Ports, Town planning)
I-Spy
London Transport
Platignum Pen Co. Ltd.
Scouse Press

Transport (*see also* Airlines, Canals, Ships, Ports, Railways)
British Aircraft Corporation Ltd.
British Overseas Airways Corporation
British Railways Board
British Transport Docks Board, Hull
British Transport Docks Board, Southampton
Canadian Pacific
Dunlop Rubber Co. Ltd.
London Transport
Manchester Ship Canal Co.
Mersey Docks and Harbour Board
Scouse Press

Underdeveloped areas
Catholic Institute for International Relations
Christian Aid
Foreign & Commonwealth Office
(Overseas Development Administration)
Overseas Development Institute Ltd.
Oxfam
United Nations Information Centre
Voluntary Committee on Overseas Aid and Development

Uranium
Commonwealth Institute

Vegetable oils
Brown and Polson Ltd.
Unilever Ltd.

Water
British Waterworks Association

Weather
Educational Productions Ltd.
Ford Motor Co. Ltd.
Meteorological Office
Diana Wyllie Ltd.

Whaling
Commonwealth Institute

Wheat

Allinson Ltd.
Association of Agriculture
Commonwealth Institute
Flour Advisory Bureau Ltd.
Hovis Ltd.
Kellogg Company of Great Britain Ltd.
McDougalls Ltd.
Pasta Information Service
Weetabix Ltd.

Wines and spirits

John Harvey and Sons Ltd.
Wine and Spirit Association of Great Britain

Wool

E. J. Arnold and Son Ltd.
British Wool Marketing Board
Harrap Bros. (Sirdar Wools) Ltd.
International Wool Secretariat

World problems

Catholic Institute for International Relations
Christian Aid
Foreign & Commonwealth Office
(Overseas Development Administration)
Overseas Development Institute Ltd.
Oxfam
United Nations Information Centre
Voluntary Committee on Overseas Aid and Development

Yeast

Distillers Co. Ltd.

[B] SOURCES AND MATERIALS IN DETAIL

For the names and addresses of suppliers abbreviated in this section, see the list at the end of the book.
*The symbol † indicates 'Head Teachers only'. An asterisk * indicates that the organisation concerned is prepared to receive enquiries from children.*

*Allinson Ltd., 210-214 Cambridge Heath Rd., London E.2
The Wholewheat story—leaflets, in a folder, about wheat and flour. Free.

Aluminium Federation, Portland House, Stag Place, London S.W.1

Specimen box, wallcharts, leaflets. Details and nominal charges available on request. Films available on free loan. List available.

E. J. Arnold & Son Ltd., Butterley St., Leeds 10

In the News series (temporarily discontinued) of charts with notes and work-cards: 'Treasure ships'; 'Olympic Games'; 'Space'. Available only while stocks last, 45s. each.
Wool Kit, 30s. Includes large wallmap, wool and cloth samples, filmstrip and notes, photo-cards and work-cards, facsimile documents and general information booklet (6 copies).

Associated Biscuits Ltd., 121 Kings Rd., Reading, Berks.

Typed notes about Huntley and Palmers, the growth of the British biscuit industry and sources of raw materials. (Associated Biscuits Ltd. is a consortium of Huntley and Palmers, W. R. Jacob and Co., Peak Frean and Co.)

Association of Agriculture, 78 Buckingham Gate, London S.W.1

Farm Study Scheme—sample studies of farms in various parts of the world, 20s. to 25s. (Listed individually under countries in Part Two.)
Bibliography of teaching material, 2s. 9d.
Handbook: *Types of farming in Great Britain,* 10s.
Booklet: *Farm studies in schools,* 8s.
Full list of all educational aids available.

Association of Jute Spinners and Manufacturers, 4 Chamber of Commerce Buildings, Dundee, Scotland

Sample box, 5s. plus 1s. postage.
Booklet: *The story of jute,* 2s. 6d.
Leaflet: *From fibre to fabric.* Teachers' notes. Free.

British Aircraft Corporation Ltd., Group Education Officer, 100 Pall Mall, London S.W.1

Brochure: *A real future in aerospace for school-leavers,* free to teachers, distributed to schools and Youth Employment Officers.

British Aluminium Co. Ltd., Education and Training Officer, Norfolk House, St. James's Square, London S.W.1

Leaflet: *Aluminium in the Highlands.* Free.

British Enkalon Ltd., Enkalon House, Regent Rd., Leicester

Booklets, diagram of manufacturing processes, samples. Free.

British Gypsum Ltd., Ferguson House, 15-17 Marylebone Rd., London N.W.1

Booklets of technical and general interest.

***British Insurance Association, Schools Project Dept., P.O. Box 538, Aldermary House, Queen St., London E.C.4**

Free booklet: *The story of British Insurance.*

British Leyland Motor Corporation Ltd., Cowley, Oxford

Catalogue of technical literature. Thirteen wallcharts specifically designed to help in instructional work in automobile engineering.

British Man-made Fibres Federation, Bridgwater House, 58 Whitworth St., Manchester 1

Leaflets, notes and small charts about a variety of man-made fibres. Free.

British Overseas Airways Corporation, The Reference Officer, P.O. Box 13, London S.W.1

Air transport study kit, containing teaching notes, colour filmstrip, route maps, timetables, work-cards, specimen tickets and one copy of *BOAC Travellers' Digest.* Available only from E.P., price 15s.
Additional copies of *BOAC Travellers' Digest* available from BOAC, price 10s. plus 1s. 6d. postage. Bulk purchase rate for teachers.
Films available on free loan from S.S.

British Petroleum Co. Ltd., Britannic House, Moor Lane, London E.C.2

Set of 19 charts, free. Film catalogue available.
Free booklets: *Energy from oil; A first look at oil; Statistical review.*

British Pottery Promotion Service Ltd., Federation House, Stoke on Trent ST4 2SA

Free leaflets and booklets. Five wallcharts, 2s. 6d. each from E.P.
Pottery picture set (9 colour prints) 15s. plus 4s. postage from C.O.I.

British Railways Board, 222 Marylebone Rd., London N.W.1

Wallcharts (as available), with colour illustration and descriptive text. Leaflets (as available), about railway passenger, motorail, shipping and hovercraft services. Copies obtainable from Public Relations Officers, British Rail, at the following regional headquarters:
Eastern Region, Hodson House, York
London Midland Region, Euston House, London N.W.1
Scottish Region, Buchanan House, Port Dundas Rd., Glasgow
Southern Region, Waterloo Station, London S.E.1
Western Region, Paddington Station, London W.2

British Ship Adoption Society, H.Q.S. Wellington, Temple Stairs, Victoria Embankment, London W.C.2

The facilities provided by the Society are only available to those schools that are already members. A leaflet outlining the aims and facilities of the Society is available, but the waiting list for membership is already a very long one.

*British Steel Corporation, Publications Room 80, 33 Grosvenor Place, London S.W.1

Free wallcharts and booklets. Films and slide sets available on free loan. Catalogue available.

British Sugar Bureau, 140 Park Lane, London W.1

Central distributors of educational material on behalf of the British Sugar Refiners.
Booklet for Juniors: *About sugar*; booklet for older pupils: *Sugar project*; handbook for teachers: *A study of sugar.*
Film: 'Sugar as energy', available on free loan from S.S.
Available from E.P.:
Wallcharts: 'Sources of sugar'; 'A sugar cane plantation'; 'Sugar beet growing'. Each 2s. 6d.
Teachers' study kit, containing handbook, two colour filmstrips, sugar samples, simulated beet and sample of cane, 20s.

British Sugar Corporation *see* British Sugar Bureau

British Trades Alphabet, East Ardsley, Wakefield, Yorkshire

An annual magazine of editorial pages, projects and competitions, covering a wide variety of topics. Intended for distribution to pupils and suitable for a wide age-range. Available in quantity for schools who wish to apply. Free.

British Transport Docks Board, Hull, P.O. Box 41, Dock Office, Queen Victoria Square, Hull, Yorkshire

Free booklets about Hull docks and unitisation. Leaflet showing trade chart for past year.

British Transport Docks Board, Southampton, Dock House, Canute Rd., Southampton SO9 1PZ

Literature about Southampton docks, and arrangements for visits (minimum age: 10 years). Apply to the Chief Docks Manager at the above address.
Films and filmstrips on Southampton docks and other transport subjects can be hired from B.T.

British Waterworks Association, The Information Officer, 34 Park St., London W.1

Booklet: *Water in your home*, 3s. from E.P.
Six wallcharts about water available from E.P. Particularly useful are No. 3, 'Irrigation', and No. 4, 'Water at work', but all six would make a valuable integrated study.
Films available for sale or hire. Full list available.

British Wool Marketing Board, Education Dept., Oak Mills, Clayton, Bradford, Yorkshire

Free booklet: *Wool, an elementary treatise.*
A selection of books on woollens, worsteds and sheep at various prices from 2s. to 5s. Wall sheets, 5s. to 9s. Twelve wool samples, 3s. 6d. Map: 'Wool in the United Kingdom', 12s. 6d.
Book of British sheep breeds, 10s.
Full details and order form available.

Brooke Bond Tea Ltd., Education Dept., Heathrow House, Bath Rd., Cranford, Hounslow, Middx.

Booklet: *The story of Tea,* 1s. (orders over 20 at 9d. each).
Wallchart 1s. 6d. Filmstrip 10s.
Booklet: *The story of Coffee,* 1s. 6d.
Map: 'The world of tea and coffee', 3s.
Films available on free loan from S.S. Full list on request.

Brown and Polson Ltd., Product Publicity Centre, 10 New Fetter Lane, London E.C.4

Recipe leaflets. Calendar, 'Vegetables in season', 10s.
Charts: 'Cooking with corn oil' and 'Cooking with corn flour', from E.P., 2s. 6d. each.

Building Societies Association, 14 Park St., London W.1

Booklets: *1000 years of Houses; 1000 years of Money; Building Societies, a two-way service.* Free.
Wallchart: 'How a Building Society works', 2s. 6d. from E.P.
Study kit containing a handbook, class sets of a booklet, wallchart and filmstrip, 7s. 6d. from E.P.

Butter Information Council, Salisbury House, London Wall, London E.C.2

Wallchart, 4s. Teachers notes, 1s. 6d. Postcards, 20 for 1s.
Information and recipe leaflets.

Canadian Pacific, Public Relations Dept., Trafalgar Square, London W.C.2

Information leaflet about railways, ships and airlines.
Filmstrips available on loan or for purchase at 15s. each.

Catholic Institute for International Relations, 38 King St., London W.C.2

Teaching kits on race and development, 10s.
C.S.E. project outline, 7s.

Cement and Concrete Association, 52 Grosvenor Gardens, London S.W.1

Full list available of priced booklets and leaflets about cement and concrete at a technical level.

Cement Marketing Co. Ltd., Portland House, Stag Place, London S.W.1

Booklet: *Cement and its uses*, up to 25 copies free.
Booklets to suit individual interests on application.
Film: 'Bond of service', available on free loan.

Central Electricity Generating Board, Sudbury House, 15 Newgate St., London E.C.1

Free booklet: *How electricity is made and transmitted.*
(*Note*: all enquiries sent to the C.E.G.B. Headquarters in London are passed for attention to the Regional Headquarters responsible for the area from which the enquiry comes.)

***Ceylon Tea Centre, 22 Regent St., Piccadilly Circus, London S.W.1**

Poster: 'The clippers'. Reference and pictorial maps of Ceylon.
Project material: 'How we get our tea'. Pictorial postcards.
Booklets: *The story of tea; How Ceylon tea is grown and marketed; All in a cup of tea.*
Full list, giving further details and list of prices, is available. Films are available on free loan.
The Ceylon Tea Centre welcomes school parties from 10.30 a.m. to 2.30 p.m. on weekdays by prior arrangement. Programme includes an introductory talk, two films in the cinema and a visit to the permanent Educational Exhibition.
Ceylon Tea Centres have also been established in Glasgow, Manchester, Leeds, Birmingham and Exeter.

Cheese Bureau, 40 Berkeley Square, London W.1

Free project book for 10-14 year old children, *Let's find out about cheese.*

Christian Aid, P.O. Box No. 1, London S.W.1

Classroom material on the need for world development, with particular emphasis on food-production and the politics of aid and trade.
Resource leaflets, study folders, schemes of work for fourth-form to sixth-form pupils.
Series of 12 pamphlets for sixth forms on overseas aid, 1s. each.
Display photographs and filmstrips with tapes available on free loan.
Films available, hire charge 20s.

Coffee Promotion Council Ltd., Botolph House, 10 Eastcheap, London E.C.3

Leaflet: *Know about coffee*; wallchart; infant frieze; Coffee ABC and game. Project card: 'Breakfast for two'.
Recipe leaflets. Free.

Commonwealth Institute, Kensington High St., London W.8

Open from 10 a.m. to 5.30 p.m. Monday to Saturday, and 2.30 p.m. to 6 p.m. on Sunday. School parties welcome—please notify the Schools Reception Centre in advance. A wide range of educational aids is available for schools and full details are provided in free leaflets, which include:
Leaflet No. 5. Maps, posters, wallcharts and other display material for sale.
Leaflet No. 6. Books, booklets and leaflets for sale.
Leaflet No. 7. Filmstrips and slides on loan.
Leaflet No. 8. Collections of study material on loan.
Leaflet No. 9. Samples of Commonwealth products for sale.
List of tape recordings available for loan.
Selected reading lists on a variety of topics.
The institute provides a schools information service for help in teaching about the Commonwealth.

continued

For teachers in Scotland:

The Commonwealth Institute, Scotland, 8 Rutland Square, Edinburgh, EH1 2AS

Operates a similar visual aids service. Catalogues and details of filmstrip and tape recording loan services are available on request.

†Copper Development Association, 55 South Audley St., London W.1

A wide range of technical publications, educational films and visual aids is available, including wallcharts on the mining and refining of copper, together with accompanying lecture notes and a selection of copper samples. A full list is available on request.

Cosmic Crayon Co. Ltd., Ampthill Rd., Bedford

Project lessons, No. 3—'Model Village'; No. 6—'Scottish Castles', 2s. each. Topic notes: No. 1 'Ships'; No. 2 'School Visits', 2s. each.

*Countryside Commission, 1 Cambridge Gate, Regents Park, London N.W.1

Series of posters, 'Our Heritage', depicting paintings of beauty spots (stocks limited). Charts about the National Parks; Exmoor, Dartmoor, Lake District, Snowdonia, Peak District, North York Moors, Pembrokeshire Coast, Yorkshire Dales.

Courtaulds Ltd., Public Relations Dept., P.O. Box 2BB, 18 Hanover Square, London W.1

Samples, charts and information, not specifically produced for geography students, but some of which is useful in this respect.

Danish Agricultural Producers Information Service, Danish Centre, 2-3 Conduit St., London W.1

Free leaflets about Danish agriculture and food.
Films available on free loan: 'Land of the Lurmark'; 'In Denmark I was born'; 'Fantasia for good food'.

De Beers Consolidated Mines Ltd., Publicity Dept., 40 Holborn Viaduct, London E.C.1

Book on famous diamonds; booklets about diamonds; wallchart. Free.

Deciduous Fruit Board, Norman House, 105 Strand, London W.C.2

Free leaflets: *Know your Cape fruit; The South African deciduous fruit industry.*
Wallchart from E.P., 2s. 6d.
Film: 'Cape of good fruit' available from S.S.

Department of Agriculture and Fisheries for Scotland, St. Andrews House, Edinburgh 1

Free booklets: *A handbook of Scotland's agriculture; Agricultural research advisory services and education in Scotland.*
Source book of agricultural education, 5s., obtainable from Oliver & Boyd Ltd., Tweeddale Court, 14 High St., Edinburgh 1. Directed mainly to teachers and students in training.

Distillers Co. Ltd., Yeast and Food Division, Crown House, Morden, Surrey

Free booklet: *The story of yeast.*

Doulton and Co. Ltd., Doulton House, Albert Embankment, London S.E.1

Charts: 'Pottery', 'Table laying', 2s. 6d. each from E.P.
Filmstrip: 'Pottery through the ages', 30s. from E.P.
Film: 'Bridging the centuries', available on free loan from Rank.

Dunlop Rubber Co. Ltd., Education Officer, 10-12 King St., London S.W.1

Booklets (maximum number per school, 20):
This is where rubber begins; Story of the wheel; Treasure in the home (story of English furniture); *The story of flight; The quickest way* (history of the development of transport); *Fort Dunlop* (story of a factory).
Sponsored aids, obtainable only from E.P.:
Filmstrip: 'The rubber plantation', 7s. 6d. (black and white). 'Rubber production on a Malayan plantation', 36s. Four wallcharts, 12s. Project booklet: *The story of rubber,* 2s.
Films available on free loan from S.S. Please give at least six weeks' notice.
Full lists of all Dunlop educational aids will be supplied on request.

Du Pont Co. (U.K.) Ltd., Press Officer (Textile Fibres), Du Pont House, 18 Breams Buildings, Fetter Lane, London E.C.4

A guide to Du Pont fibres, leaflets, chart and samples. Free.

Dutch Dairy Bureau, Dutch House, 307-8 High Holborn, London W.C.1
Leaflets for children. Colour films (16 mm)—list on request.
Wallchart: 'How Cheese is Made', 2s. 6d. from E.P.

Educational Productions Ltd., East Ardsley, Wakefield, Yorkshire
Five wallcharts about the weather, with teachers' notes, 52s. 6d. per set.
Two filmstrips: 'Air routes of the World', (i) London-Karachi, (ii) Karachi-Tokyo, each 36s. 'Air Age World Map' (4 charts), 30 x 40 in., per set 30s.
Full details of all educational aids, including kits, sponsored material, filmstrips, charts, project material, etc. can be obtained on application.

*Elders and Fyffes Ltd., 15 Stratton St., Piccadilly, London W.1
Free booklet: *The banana story.* Black-and-white filmstrip, 21s. from E.P. Wallchart, 2s. 6d. from E.P.

Electricity Council Marketing Dept., Trafalgar Buildings, 1 Charing Cross, London S.W.1
Advertising literature. Booklet: *Electricity and you.*
Films available on free loan, list available.

English China Clays Group, John Keay House, St. Austell, Cornwall
Booklets about the China clay industry.

Esso Petroleum Co. Ltd., Public Affairs Dept., Educational Services, Victoria St., London S.W.1

Wallcharts and booklets on a wide variety of geographical and technical themes. Full list available.
Esso Geography Studies, 15s. each, published by Longman and available through normal distributors. Twenty-five studies of the geography of Great Britain, each comprising 6 colour slides and notes. (For individual titles, see British Isles entries in Part 2.)

*European Community Information Service, 23 Chesham St., London S.W.1

Full list of publications available, including booklets, maps, leaflets, films, filmstrips, display photographs and travelling exhibitions. Useful information on Common Market affairs.

Farmer and Stockbreeder Publications Ltd., Publishing Dept., Dorset House, Stamford St., London S.E.1

Picture charts on a large variety of animals, 9d. each, except 'Farm stock of Britain', 1s. 3d.

†Fish Education Service of the White Fish Authority, Dept. BTA. 69, 20 Cannon St., London E.C.4

Wallchart 3s. 6d. Recipe booklet–free.
Film: 'Trawler Boy'. Free loan on request of Head Teachers.

Flaxspinners and Manufacturers Association of Great Britain, 4 Chamber of Commerce Buildings, Dundee, Scotland

Leaflet: *The story of linen.* Send stamp.
Visual Aid box, 10s. 6d.

Flour Advisory Bureau Ltd., 21 Arlington St., London, S.W.1

Pamphlet: *The milling of wheat.* Booklet: *The why and where of wheat.*
Leaflet: *Know your wheat grain.* Sample card. Free.
Filmstrips from E.F.V.A.

Ford Motor Co. Ltd., Advertising and Sales Promotion Dept., Cranes Farm Rd., Basildon, Essex

Supply of posters, containing 30 coloured photographs of clouds, with captions describing the cloud and giving a brief guide to the weather likely to be associated with each type of cloud.
'Facts from Ford'–pack of excellent teaching materials.

*Foreign & Commonwealth Office (Overseas Development Administration), Information Dept., 119 Eland House, Stag Place, London S.W.1

A wide range of material on overseas development and aid, including fact sheets, booklets, displays and reports. A full list, with prices, is available on request.
Films are available from C.O.I. (C.F.L.).

*Forestry Commission, 25 Saville Row, London W.1
Booklets: *Starting a school forest; Forestry and the town school; A guide to home-grown timbers.*
Full list entitled 'Schools and Forestry' available, giving details of filmstrips, films, samples, wallcharts and nature trails. A booklist is also supplied.
Note: items for loan in the Forestry Commission list are not available to student teachers.

*J. S. Fry and Sons Ltd., Somerdale, Bristol BS99 7HW
Free leaflets about J. S. Fry and Sons and the manufacture of chocolate.

Furness Withy and Co. Ltd., 56 Leadenhall St., London E.C.3
No material specially produced for schools, but this organisation, representing several shipping lines, is willing to send a selection of publications to teachers with a genuine interest.

Gas Council Film Library, 6-7 Great Chapel St., London W.1
Film library service. Films available on free loan.
Catalogue on request. Of particular value: 'The methane story'; 'Saharan venture': 'Sea-born treasure'.

Glass Manufacturers Federation, 19 Portland Place, London W.1
Folder of leaflets, free on request.

*Goodyear Tyre and Rubber Co. (Great Britain) Ltd., Public Relations Division, Bushbury, Wolverhampton, Staffs.
Free booklets: *Story of the tyre; This is Goodyear.*

Gypsum Products Development Association, P.O. Box 321, 15 Marylebone Rd., London N.W.1
Free booklet: *Gypsum today.* Wallchart, 3s. 6d.

Hamish Hamilton Ltd., 90 Great Russell St., London W.C.1
Trailblazers project folders on Rivers, Forests, Markets and Farming. (Future titles will include Ports, Mining and Space.) Contain work–cards, information leaflets and pictures. Each 25s.

*Harrap Bros. (Sirdar Wools) Ltd., Dept. B.T.A., Bective Mills, Wakefield, Yorkshire
Booklets: *From fleece to fashion; Growing up in Sirdar.* Small sample card. Free. Wallchart, 3s. 6d., from E.P.

John Harvey and Sons Ltd., Publicity Dept., 12 Denmark St., Bristol 1
Booklet: *A guide to wine*; Map leaflets about wine-growing areas in France. Poster about wines of Spain.

*Heinz Home Cookery Service, Greater London House, Hampstead Rd., London N.W.1
Cookery information. Booklets: *How food is canned; Heinz and the world of food.*

†Herring Industry Board, 1 Glenfinlas St., Edinburgh 3

Booklets: *The herring book; The story of the herring.* Single copies free, then 6d. each.
List of filmstrips available for purchase; particularly useful: 'Our herring industry', 6s.

*Hovis Ltd., 'Wheatgerm', Macclesfield, Cheshire

Booklet: *Looking at wheat the Hovis way* (contains specimens).
Wallchart: 'The wheat plant', available from E.P.

†Hudson's Bay Co., Beaver House, Great Trinity Lane, London E.C.4

Free booklet: *A brief history of the Hudson's Bay Co.*

I.C.I. Ltd., Imperial Chemical House, Millbank, London S.W.1

Large range of books available from publishers and booksellers. Full list available. Example: *Limestone as a raw material in industry*, 10s. 6d.
Books and leaflets from the above address, e.g. *Magadi, the Crystal Lake.*
Films available on free loan from the I.C.I. Film Library, address as above.

*I.C.I. Fibres Ltd., 68 Knightsbridge, London S.W.1

Free booklet: *Facts about I.C.I. synthetic fibres.*

I.C.I. Plastics Division, Publicity and Information Services Dept., Welwyn Garden City, Herts.

Leaflets: *I.C.I. and the plastics industry; Buying and selling plastics.* Index of technical literature available.
(*Note:* Much of this material is more suited to science.)

*Industrial Diamond Information Bureau, 7 Rolls Buildings, Fetter Lane, London E.C.4

Wide range of information, including wallcharts, booklets, filmstrips, slides and films. Full publications list available.

Inland Waterways Association Ltd., 114 Regents Park Rd., London N.W.1

Free information about all aspects of inland waterways.

International Wool Secretariat, Dept. of Education and Training, U.K. Branch, Wool House, Carlton Gardens, London S.W.1

Black-and-white filmstrip: 'Wool cloth manufacture', 9s.
Cloth sample book, 5s. 6d. Booklets about wool, 1s. to 1s. 6d. Scottish wool cloth sample book, 7s. 6d. Wallchart display card, 6s. 6d. Booklets about knitting for the Primary age-group.
Films available on free loan. Full list of all educational aids available on request.

*Irish Linen Guild, Morley House, 314 Regent St., London W.1

Free booklets: *Living with Irish linen; Making and using Irish linen.*
Wallcharts: 'The story of Irish linen', 'Living with Irish linen', 5s. per set from E.P.

continued

Educational wall frames, available only on receipt of an official requisition, 10s. for postage and packing.
Educational box, including filmstrip, samples and teachers' guide, 10s. for postage and packing. Official requisition only.
Films available on free loan: 'To last a lifetime'; 'Looking in on Irish linen'.

*I-Spy, Big Chief I-Spy, Wigwam by the Water, 161 Queen Victoria St., London E.C.4

I-Spy series of booklets. Wide range of titles. Full list available on request. Some particularly useful titles:
Seaside; Country; Sights of London; On the farm; History; In the street; On the road; The sky. 1s. 6d. each.
On a car journey; On a train journey; Abroad; Archaeology; On the sea shore. 2s. 6d. each.
Books obtained from the above address may be purchased at a special discount of 25%, and re-sold to pupils at the normal price so that school funds may benefit.
'I-Spy' Competition Sheets for towns are available in small quantities. The competitions are now closed, but the questionnaires make useful itineraries for the exploration of towns. Leaflets available for Liverpool, Hastings, York, Chester, Bristol, Shrewsbury and Bath.

*Jaffa Information Bureau, Benton and Bowles Ltd., 197 Knightsbridge, London S.W.7

Free booklet: *Israel, land of Jaffas.*

Kellogg Company of Great Britain Ltd., P.O. Box 278, Manchester

Free booklet: *The grains are great foods.*
Wallchart: 'Story of Maize', from E.P., 3s. 6d.
Colour filmstrip: 'The story of corn', from E.P., 36s.

*Kraft Foods Ltd., Regina House, Old Marylebone Rd., London N.W.1

Free wallchart and teachers' notes about processed cheese.

Lead Development Association, 34 Berkeley Square, London W.1

Notes about lead for students, wallchart. Free.

*Leather Institute, Leather Trade House, 9 St. Thomas St., London S.E.1

Leaflet: *The wonderful story of leather.* Notes about the location of the leather industry in Great Britain, and curing skins. Free.
Films: 'L for leather' and 'Traders in leather' available on free loan from S.S.

*Libby McNeill and Libby Ltd., 396 Kenton Rd., Kenton, Harrow, Middx.

Wallchart: 'Peach growing in California', from E.P., 2s. 6d.

†London Transport, Publicity Officer, 280 Marylebone Rd., London N.W.1

Maps, timetables, posters, postcards and booklets available for purchase. Full list on request.

Joseph Lucas Ltd., Central Advertising Dept., Great King St., Birmingham 19

Free booklets and leaflets about the Lucas Company, Ignition (Fire), the story of Light and the story of Power.

Lullingstone Silk Farm Ltd., Ayot House, Ayot St. Lawrence, Herts.

Admission to silk farm: children 2s. 6d., adults 4s. Special rates for school parties. Souvenirs and literature available.

J. Lyons and Co., Ltd., (Dept. T8), Cadby Hall, London W.14

Free booklets: *The story of coffee; The story of tea.*
Study folders: 'About tea', 'About coffee'. Free.

***McDougalls Ltd., Dept. B.T.A., Station Rd., Daybrook, Nottingham**

Leaflet: *The story of a bag of flour*, free. Maximum 50 per school.
Set of project stamps, free. Maximum 50 per school.
Wallchart: 'How self-raising flour is made', 7s. 6d.
School project box, 3s. 6d. Cookery books and posters.

Manchester Ship Canal Co., Public Relations Dept., Ship Canal House, King St., Manchester M2 4WX

Free booklets: *The Port of Manchester; Barton Aqueduct.*
Notes on the Port of Manchester.
For teachers only, free pamphlet: *Resolution and achievement.*

Mersey Docks and Harbour Board, Information Officer, Pier Head, Liverpool 3 1BZ

Free booklets about the docks. Map folder of the port.

Meteorological Office Library, London Rd., Bracknell, Berks.

Library service of books, pamphlets, slides and black-and-white photographs. Reference list available.
Daily weather report service for schools and colleges during term. Subscription 112s. 6d. per annum.
Wide range of publications available at various prices.
Films available for hire from Rank and from Services Kinema Corporation, Chalfont Grove, Gerrard's Cross, Buckinghamshire.

***Milk Marketing Board, Thames Ditton, Surrey**

Information leaflets and brochures covering all aspects of milk production, transport and marketing. Films and specialist speakers also available.

Ministry of Agriculture, Fisheries and Food, The Library, 3 Whitehall Place, London S.W.1

Picture sets and coloured photo-posters available on free loan. Four food group posters, free. Long list of potential sources of information.

Monsanto Textiles Ltd., Public Relations Dept., 10-18 Victoria St., London S.W.1

Free folder of leaflets, diagrams and samples dealing with the manufacture of Acrilan and Blue 'C' Nylon.

National Coal Board, Public Relations Dept., Hobart House, Grosvenor Place, London S.W.1

A large selection of booklets, leaflets and wallcharts about the British coal industry. Films available on free loan. Catalogue available.

National Dairy Council, National Dairy Centre, John Princes St., London W.1

A catalogue of educational material is available free of charge. This illustrates all the material produced by the Council, including booklets, charts, postcards, teachers' notes about milk, butter, cream, cheese and cows. Filmstrips and careers literature also available.

National Farmers Union, Information Division, Agriculture House, Knightsbridge, London S.W.1

Free leaflets: *Talking about agriculture; Talking about horticulture; Food and farm facts* (intended specially for children); *Farming as a career* (for careers masters); *Basic statistics.*
Economics leaflet: *Eaten your travel allowance yet?*
Film: 'Look to the land', available on free loan and suitable for Secondary pupils.
Farm visits can usually be arranged for parties of children over the age of 11 years.

*National Federation of Clay Industries, 30 Gordon St., London W.C.1

Free leaflet: *How bricks are made.*
Films: 'Bricks and brick masonry'; 'Roofs of clay'. Available on free loan from Rank.

*National Federation of Fruit and Potato Trades Ltd., Russell Chambers, Covent Garden, London W.C.2

Calendar chart about fruit and vegetables and a free booklet.

Natural Rubber Producers Research Association, 56 Tewin Rd., Welwyn Garden City, Herts.

Free booklet: *The story of natural rubber and the people of the plantations.*
Film: 'Naturally it's rubber', available on free loan from S.S.

Nestlé Co. Ltd., St. Georges House, Croydon, Surrey

Free booklet: *The story of coffee.*
Filmstrip about cocoa, milk, sugar and the manufacture of milk chocolate, 7s. 6d.
Wallchart for teachers, 1s. 6d. postage and packing.

New Zealand Dairy Board, St. Olaf House, Tooley St., London S.E.1

Free booklets: *Dairy farming in New Zealand; The manufacture of dairy produce*—for senior students only.

*New Zealand Shipping Co. Ltd., P. & O. Building, Leadenhall St., London E.C.3

Free notes and picture folder.

Norvic Shoe Co. Ltd., St. Georges Plain, Norwich NOR 32A

Free leaflets about shoe-manufacture and leather.

*Outspan Organisation, 6 Henrietta St., Covent Garden, London W.C.2

Free booklets: *The story of Outspan; Food value of citrus fruits.*
Wallchart from E.P., 2s. 6d. Films available on free loan from S.S.

†Overseas Development Institute Ltd., 160 Piccadilly, London W.1

Wide range of publications available. Full list on request. Particularly useful: *Opting for development,* 8s. 6d.; *World III,* 12s. 6d.; *Rich world, poor world,* 3s. 6d.
All book orders should be sent through Research Publications Services Ltd., 11 Nelson Rd., Greenwich, London S.E.10.

Oxfam, 274 Banbury Rd., Oxford

Free information sheets and leaflets about Oxfam, general hunger problems, health, refugees, development, etc.
Primary school wallcharts, project folders, photographs, and books for book corners and libraries.
Wide range of aids for Secondary schools. Full details and prices available on request.
Films available for loan, 10s. each. List on request.

Pan Am Publications, Pan American Airways, 193 Piccadilly, London W.1

Catalogue of publications available.

Pasta Information Service, 15 Lots Rd., London S.W.10

Illustrated project folder: 'Meet the Pasta Family'.
Pasta shapes chart. Visual-aid samples and teaching notes.
Wallchart from E.P., 2s. 6d.
Film: 'Spaghetti varieties', available on free loan from S.S.

Petroleum Information Bureau, 4 Brook St., London W.1

Free booklets about prospecting, drilling, refining, transport and distribution, oil-producing areas. Statistics. Booklet about natural gas. Film catalogue available from Petroleum Film Bureau at the above address.

Pilkington Bros., Ltd., Glass Manufacturers, St. Helens, Lancs.

Free booklets: *Glass making; Glass, the Pilkington story.*

*Platignum Pen Co. Ltd., Schools Division, Platignum House, Six Hills Way, Stevenage, Herts.

Discovery Series of project folders of environmental studies, based on Stevenage as an example, with work sheets for pupils, 3s. 6d. each. Titles available are: Geography, History, Agriculture, Industry, Community and Social.

†John Player and Sons, Castle Tobacco Factory, Nottingham

Notes about J. Player and Sons. Typescript: 'From Plantation to Players'. Map: 'World sources of tobacco'.

†Port of Bristol Authority, Queen Square, Bristol 1

Information about Britain's largest municipal port, comprising City Docks, Avonmouth Docks in Bristol and Portishead Dock in Somerset, and including details of a proposed development, 'Bristol Western Docks Scheme'. Information will only be sent to schools, colleges and libraries etc., and not to individuals.

Port of London Authority, Public Relations Officer, P.O. Box 242, Trinity Square, London E.C.3

Booklet: *Ships and cargoes*, 1s. Notes on the Port of London, 3s.
Wall frieze of 6 charts, 10s.
Monthly magazine, subscription 30s. per annum.
Postcards, 6d. each. List of reference books.
Filmstrips: 'London docks', 22s. 6d. from C.G. 'Port of London', 20s. from Hulton.
Films: 'Port of London' available from the Central Publications Film Library; 'Container Port' available from B.T.

Potato Marketing Board, 50 Hans Crescent, London S.W.1

Wallchart from E.P. 2s. 6d. 'How potatoes are grown'.
Booklet: *The history of the potato*, 3s. 6d. from E.P.
Pupils' leaflet, 1s. per dozen from E.P.
Films available on free loan: 'A crop of problems'; 'A crop for all seasons'.
Full list of all educational aids on request.

*Raleigh Industries Ltd., Dept. B.T.A., Nottingham

Small folder to stimulate interest in cycle touring. Includes simple map work. Free.

Reed Paper Group Ltd., Education Services Dept., Reed House, Piccadilly, London W1A 1EJ

Wallchart about papermaking. Sample set of papers.
Booklet: *Paper, material of 1000 uses.* Teachers' notes.
Filmstrip: 'Paper and its production', 25s. or 62s. 6d. complete with record. Black-and-white film: 'Mill by the Medway' available on free loan. The filmstrip and record are also available for loan.

*Rice Council, 87 Notting Hill Gate, London W.11

Teachers' kit. Student teachers' pack. Leaflets: *Facts about American rice; History of American rice; Seven thousand varieties of rice.*
Film: 'The story of American rice', available on free loan from S.S. and Rank.

Roan Selection Trust Group, 1 Noble St., London E.C.2

Booklets on the mining and uses of copper. Wallchart with pictures and text descriptions of the mining and metallurgical treatment of copper in Zambia. Free.

*James Robertson and Sons Ltd., Golden Shred Works, 138 Bromley Rd., London S.E.6

Booklet: *The story of Golden Shred.* Wallchart: 'Some edible fruits'. Free.

Rowntree and Co. Ltd., The Cocoa Works, York YO1 1XY

Samples of raw materials with illustrated booklets.
Shop set of dummy packs for school 'shops'.

*Royal National Lifeboat Institution, Public Relations Officer, Lifeboat House, 42 Grosvenor Gardens, London S.W.1

General notes; school posters; booklet: *The story of the lifeboat,* 3s. Three wallcharts.
(Though the R.N.L.I. is willing for children to write requesting material, teachers should remember that the Institution is supported entirely by voluntary contributions.)

Scottish Association of Young Farmers Clubs, 11 Rutland St., Edinburgh

Information on the general practice of farming in Scotland.

Scouse Press, 4 Windermere Terrace, Liverpool 8

Liverpool Packet No. 2, 'The growth of a city', a collection of maps, plans and views, 15s.; No. 3, 'Liverpool transport through the ages', 15s.; No. 4, 'The Liverpool–Manchester Railway', 21s. Postage extra.
Full list of publications available.

Shell International Petroleum Co. Ltd., Shell Centre, London S.E.1

Wallcharts, booklets, slides, filmstrips and project notes about oil and related subjects, supplied free or on free loan to a school address only. Films are available on free loan (postage fee), to schools through P.F.B. 8 mm loops available from Gateway. Free catalogues are sent to school addresses only, at the request of a member of staff. Students training as teachers may request a booklet on oil for their own information, but should obtain wallcharts and other teaching aids through the schools where they do practice-teaching.

Shipbuilders and Repairers National Association, 21 Grosvenor Place, London S.W.1

Background notes on the British shipbuilding industry.
Film: 'Ships for the world: built in Britain', available on free loan from C.O.I. (C.F.L.).

***Silk Educational Service, 10 Cliffe Rd., Barton-on-Sea, Hants.**

Sets of educational material: *Set A,* for teachers of geography and natural history, 7s. This set includes leaflets, brochures, photographs and specimens. *Set C* is intended for pupils, price 2s. A full list of all educational material and an order form can be obtained on request.
Note: The Silk Educational Service now handles all educational material on behalf of the Silk Centre, Sewing Silks Ltd., and H. T. Gaddum Ltd.

***Smedleys Ltd., Education Dept., Whyteleafe, Surrey**

Free booklet: *Smedleys and the foods we eat.* Class sets supplied on request.

†Smiths Industries Ltd., Aviation Division, Bishops Cleeve, Cheltenham, Glos.

Free booklets: *Smiths Industries Aviation Division; About Smiths.*

Southern Postal and Duplicating Services, 63 Old Town, Clapham, London S.W.4

Free booklets: *Apple time in New Zealand; Cape fruit.*

A. Strauss and Co. Ltd., Plantation House, Mincing Lane, London E.C.3

Free map: 'The geography of lead'.

Textile Council, Education and Information Dept., 3 Alberton St., Manchester 3

Filmstrip: 'Cotton growing', 10s. Sample box, 7s. 6d.
Fabric folder, 2s. Leaflet: *Growing cotton plants,* 1s.
Wallcharts, 2s., 2s. 6d., 3s. 6d. Booklet: *Cotton guide,* 2s. 6d.
Films available on free loan. Full list available on request.

Timber Research and Development Association, Hughenden Valley, High Wycombe, Bucks.

Wall map of world forest regions and commercial timbers, 15s. (paper) or 40s. (cloth on rollers).
Nineteen wallcharts dealing with the growth, uses, and processing of timber, plywood etc., 9s. each .
Forty-four labelled samples, 2s. 6d. each or 80s. the set.
Free list of publications and teaching aids available on request.

Topic Records Ltd., 27 Nassington Rd., London N.W.3

Suppliers of recorded folk music suitable for use in school lessons.
Catalogue available. Free leaflet: *Folk music and the geography lesson* (reprint of an article).

Town and Country Planning Association, The Planning Centre, 28 King St., London W.C.2

Limited information for schools about new and expanded towns. The Association will provide help in response to genuine enquiries and will provide bibliographies, etc. for teachers and students.
Subscription to the *Journal of Town and Country Planning,* 37s. 6d. per year.

Trinity House Lighthouse Service, Trinity House, Tower Hill, London E.C.3

Free booklet: *The History of Trinity House.* Cross-section of Eddystone Lighthouse. Map showing the position of lighthouses and lightships, free (one only per school or class). List of lighthouses open to the public.

Unilever Ltd., Film and Education Dept., Information Division, Unilever House, Blackfriars, London E.C.4

Free booklets about detergents, vegetable oils and fats. Free wallcharts about soap, margarine, vegetable oil, fish.
Filmstrips about soap, margarine and vegetable oil.
Films available on free loan. Catalogue available on request, together with order form.

Union-Castle Mail Steamship Co. Ltd., Cayzer House, 2 St. Mary Axe, London E.C.3

Fact sheets and pamphlets about the Union-Castle Line.

†United Kingdom Atomic Energy Authority, Public Relations Branch, Library and Information Centre, 11 Charles II St., London S.W.1

Pamphlets and photographs, free. Films on free loan. List available.

†United Nations Information Centre, 14-15 Stratford Place, London W.1

Leaflets about the United Nations and subsidiary organisations.
Wallcharts according to current availability. Filmstrip and film catalogues available.
Note: No material available about continents or specific countries.

Van Den Berghs Ltd., Dorset Rise, Kildare House, London E.C.4

Wallcharts: 'Margarine manufacture'; 'Raw materials for margarine'.

Volkswagen Motors Ltd., Volkswagen House, Brighton Rd., Purley, Surrey

Leaflets about car-production. Large display pictures.
Typed notes with reference to the Volkswagen factories in Germany.
Book: *Small Wonder: the amazing story of the V.W.,* by W. H. Nelson, available for purchase. (Published by Little, Brown & Co., Toronto.)

*Voluntary Committee on Overseas Aid and Development, The Education Officer, 69 Victoria St., London S.W.1

This organisation promotes educational programmes for the increased awareness of the problems of the developing countries. The organisation co-ordinates the educational work of:
The Catholic Institute for International Relations; Christian Aid; Oxfam; Overseas Development Institute; Freedom from Hunger Campaign; War on Want; Save the Children Fund; United Nations Association.
A very wide range of teaching aids is available, including fact sheets, booklets, charts, kits, filmstrips, etc. Full lists are available.
Pamphlets published by the Committee include *Impact*, an annual survey of United Kingdom Examination papers and syllabuses; and *Student teachers look at the developing countries*, a review of College of Education syllabuses.

*Josiah Wedgwood and Sons Ltd., Barlaston, Stoke-on-Trent, Staffs.

Educational sets, including raw materials and partly finished pieces, 30s. each. Postcards, 6d. each. Illustrated brochures, free. Folder map: 'The way to Wedgwood', free. Booklet: *Know your Wedgwood*, free. Booklet: *The making of Wedgwood*, 2s. Films available on free loan from Rank.

Weetabix Ltd., Weetabix Mills, Burton Latimer, Kettering, Northants.

Free pamphlet for children: *How wheat becomes your breakfast cereal.* Chart (for teachers only): 'The romantic story of wheat'.

Wiggins Teape and Co. Ltd., Head Office, Gateway House, 1 Watling St., London E.C.4

Leaflet about papermaking. Book of sample papers.
Booklet: *Paper in this modern world.* Library kit for making paper.
Films available on free loan: 'Paper chain'; 'Gateway to the future'.

†W. D. & H. O. Wills, Public Relations Dept., Bedminster, Bristol BS99 7UJ

Free booklets: *Pacemakers in tobacco; Tobacco, its history, culture and manufacture.*
Free map of the main tobacco-growing areas of the world.

Wine and Spirit Association of Great Britain, Education Dept., 3 Lovat Lane, London E.C.3

Overhead projector transparencies available for sale.
Films for hire. Book list. Full list of aids available.

Diana Wyllie Ltd., 3 Park Rd., Baker St., London N.W.1

Slides and filmstrips about meteorology, including cloud formation, weather stations, meteorological instruments, etc.
Catalogue and price list available.

REGIONAL LIST OF SOURCES AND MATERIALS

This list also includes, placed for covenience within the alphabetical sequence, entries for Climate, meteorology and vegetation, Mathematical geography, and Physical geography.

min. = *minute*	f. = *frames*	*h.* = *hire fee*
sec. = *second*	sl. = *slides*	b/w = *black and white*

The symbol † indicates 'Head Teachers only'. For the names and addresses of suppliers abbreviated in this Part, see the list at the end of the book.

ADEN

see Southern Yemen

AFRICA

see also South Africa

The African Bureau, 38 Great Russell St., London S.W.1
Small bureau interested in the political and cultural concerns of Central and South Africa; bi-monthly *African Digest*, 25s. per annum.
Coffee Promotion Council, 10 Eastcheap, London E.C.3
Wallfrieze for infants. World map of production and African processes. Both free.
Timber Research and Development Association
Timbers of West Africa, 8s.
Petroleum Information Bureau
Free booklet: *Oil in Africa.*

Maps–topographic

Map lists for Africa and each territory from Stanfords.

Atlases

Modern College Atlas for Africa		17s. 6d.	Philip
School Atlas for East Africa		5s.	Philip
East African Modern School Atlas		16s.	Philip
Supplements only	per ten	20s.	Philip
Atlas for Eastern Nigeria		5s.	Philip
New School Atlas–West African edition		13s.	Philip
Supplements only	per ten	20s.	Philip
Oxford Regional Economic Atlas of the Middle East and N. Africa		70s.	O.U.P.

Films

Oil in the Sahara	19 min.	free loan	P.F.B.
Continent of Africa	19 min.	*h.* 40s.	Rank
Children of Africa (1966) b/w	20 min.	hire fee	S.S.

continued

Life in an oasis (N. Africa)	b/w	10 min.	*h.* 15s.	Gateway
Tropical Africa		22 min.	*h.* 50s.	Gateway

Film cassettes (8 mm)

Cultural groups			
Town life			
Village life			
Tropical products			
Crossing Sahara	all 3-4 min.	each 69s.	Gateway

Filmstrips

Village life		47 f.	42s.	Gateway
Drift to the town		40 f.	42s.	Gateway
Town life		32 f.	39s. 6d.	Gateway
Congo handclasp		55 f.	*h.* 2s. 6d.	B. & F.
Life among the Arabs			44s.	C.G.
Ports of West Africa			44s.	C.G.
Farmers of Africa	b/w		23s.	C.G.
Natives of Africa	b/w		23s.	C.G.
Congo Basin	b/w		23s.	C.G.
North West Africa	b/w		23s.	C.G.
Peoples of Africa	b/w		20s.	Hulton
Bushmen of Africa	b/w		20s.	Hulton

Also long list from C.I. and V.C.O.A.D.–free.
List from E.F.V.A., 3s.

Wallmaps and charts

Africa	30 x 40 in.	10s. 6d.	Pictorial
W. African countries (5 charts)	15 x 20 in.	per set 10s. 6d.	Pictorial
Africa (with notes)		10s. 6d.	C.I.
Wallmap	39 x 31 in.	5s. 6d.	C.I.
Changing map of Africa	13 x 24 in.	1s. 2d.	C.I.
Kariba Dam	postcard	8d.	C.I.
Physical and economic (2-sided)	48 x 36 in.	85s.	Macmillan
Physical (Westermann)	65 x 73 in.	175s.	Grant
Physical	47 x 68 in.	110s.	Philip
Graphic relief	34 x 42 in.	42s.	Philip
Political	47 x 68 in.	110s.	Philip
Wall Atlas (5 maps)–Relief, climate, vegetation, political, population	43 x 33 in.	per set 170s. per map 42s.	Philip
Central Africa–Physical and communications	37 x 46 in.	44s.	Philip
Colonial development (Westermann)	71 x 79 in.	185s.	Grant

continued

N. Africa–Physical			
(Westermann)	102 x 65 in.	185s.	Grant
Black outline	30 x 40 in.	12s. 6d.	Hulton

ALGERIA

Barclays Bank

Economic report (ask for whole set)

ARGENTINA

Argentine Embassy, 9 Wilton Crescent, London S.W.1

Excellent map and colour brochures on regions of the country. Profile booklet, fact sheets, posters, some booklets in Spanish.

Barclays Bank

Economic report (ask for whole set)

Atlas

Atlas Medio Universal y de la República Argentina, from Barmerlea Book Sales Ltd., Annandale, North End Rd., London N.W.11

Films

Argentina	b/w	11 min.	*h.* 12s. 6d.	Rank
Horsemen of the Pampas	b/w	20 min.	*h.* 20s.	Rank
Argentina		21 min.	*h.* 30s.	Foundation

Filmstrips

Argentina	b/w	40 f.	16s.	V.I.S.
Children of the Pampas		41 f.	55s. 3d.	McGraw
Argentina		43 f.	57s. 6d.	McGraw
Horsemen of the Pampas	b/w	51 f.	18s.	Rank
Argentine meat	b/w		20s.	Hulton
Argentine village	b/w		20s.	Hulton

Slides

La Plata countries	25sl.	per set 85s.	Grant

ASIA

Timber Research and Development Association

Booklet: *Timbers of South East Asia,* 8s.

Atlas

Atlas of South-east Asia	50s.	Macmillan

Films

Asia		16 min.	*h.* 50s.	Gateway
Children of Asia	b/w	15 min.	hire fee	S.S.
Mekong		25 min.	free loan	P.F.B.

Filmstrips

Farmers of Asia	b/w		23s.	C.G.
Community development		28f.	40s.	Foundation

Slides

S.E. Asia	25sl.	per set	85s.	Grant

Wallmaps and charts

General map	30 x 40 in.		5s. 6d.	C.I.
S. E. Asia (3 charts)	37 x 25 in.	per set	22s. 6d.	E.P.
Black washable 2-sided (Africa/Asia)	54 x 42 in.		110s.	Grant
Political	71 x 66 in.		110s.	Philip
Physical	71 x 66 in.		110s.	Philip
S.E. Asia–Political	45 x 36 in.		44s.	Philip
Graphic relief	34 x 42 in.		42s.	Philip
Contoured outline (paper)	41 x 32 in.		10s.	Philip
Wall Atlas (5 maps)–Relief, climate, vegetation, political, population	43 x 33 in.	per set	170s.	
		per map	42s.	Philip
Geography of Asia (16 charts)	30 x 40 in.	per set	80s.	E.P.
Asia (Eurasia)–Physical (Westermann)	84 x 90 in.		175s.	Grant
S.E. Asia–Political (paper)	13 x 24 in.		1s. 1d.	C.I.
Asia (paper)	30 x 40 in.		3s. 6d.	P. & T.
Outline	7 x 5 in.	per 25	3s. 6d.	E. J. Arnold

AUSTRALIA–GENERAL

Australia and New Zealand Bank, 71 Cornhill, London E.C.3
Booklets and books on economic development and prospects.

†Australian Migration Information Service, Bank of New South Wales, Television House, 14 Kingsway, London W.C.2
Booklets on currency, stamps and exploration. Limited supplies.

Australian News and Information Service, Canberra House, Maltravers St., Strand, London W.C.2
Very good. Picture booklets on State capital cities, Canberra, animals and flowers.
Films and slide catalogue available.
Excellent filmstrips catalogue. Colour slides on loan.
Official handbook.

Atlases

Atlas of Australian Resources	210s. from Stanfords Separate plates each 5s.	
Modern Commonwealth Atlas	(price on application)	Cochrane
Primary Atlas for Australian Schools	(price on application)	Cochrane
Oxford Australian Atlas	22s.	O.U.P.

Maps–topographic

1:1,000,000; 1:250,000; 1:50,000 available from Stanfords

Films

Back of beyond	b/w	65 min.	free loan	P.F.B.
Gold mining in Australia	b/w	12 min.	*h.*15s.	Gateway
Australia	b/w	11 min.	*h.*12s. 6d.	Rank

Filmstrips

Farmers of Australasia	b/w		23s.	C.G.
Darwin to Adelaide			42s.	C.G.
Everyday life in Australia: Agriculture; Industry; Aborigines			each 36s.	E.P.
General survey	b/w		21s.	E.P.
Victoria and Tasmania	b/w		21s.	E.P.
Eastern Australia	b/w		21s.	E.P.
Central lowlands	b/w		21s.	E.P.
Swanland and South Australia	b/w		21s.	E.P.
Monsoonlands and the arid interior	b/w		21s.	E.P.
Map strip	b/w		10s.	V.I.S.
Pastoral industries	b/w		15s.	V.I.S.
Crops and forestry	b/w		15s.	V.I.S.
Minerals and industry	b/w		15s.	V.I.S.
Wild life and vegetation	b/w		15s.	V.I.S.
Sheep ranch country	b/w	51f.	18s.	Rank
Sheep farming			42s.	C.G.
Irrigation farming			40s.	Foundation

Slides

Australia	25sl.	per set 85s.	Grant

Long list for free loan from C.I.

Wallmaps and charts

Physical—Australia & Oceania (Westermann)	82 x 70 in.	175s.	Grant
Political	69 x 48 in.	110s.	Philip
Physical	70 x 48 in.	110s.	Philip
Physical and communications (Australia & E. Asia)	45 x 37 in.	44s.	Philip
Graphic relief	34 x 42 in.	42s.	Philip
Wall Atlas—Australia & New Zealand (5 maps)—Relief, climate, vegetation, political, population	43 x 33 in.	per set 170s. per map 42s.	Philip
Black outline (paper)	30 x 40 in.	12s. 6d.	Hulton

continued

Primary industry (9 charts)	24 x 36 in.	per set	12s. 6d.	C.I.
Flying doctor	20 x 30 in.		3s. 9d.	C.I.
Topographic (paper)	30 x 40 in.		5s. 6d.	C.I.
Political	43 x 33 in.		11s.	C.I.
Australian sheep farming	30 x 40 in.		8s. 6d.	Pictorial
Australia (4 charts)	15 x 20 in.	per set	10s. 6d.	Pictorial
Outline maps	7 x 5 in.	per 25	3s. 6d.	E. J. Arnold

AUSTRALIA–NEW SOUTH WALES

New South Wales Government Office, 56 Strand, London W.C.2
Handbook, schools pamphlet, 35 mm slides.
Films for free loan: 'The New Australians', 'City of Millions' (Sydney)

Association of Agriculture

Sheep-Wheat farm study–maps, diagrams and text with photographs, 25s.

Films

September spring (Snowy Mountains)	19 min.		free loan	P.F.B.

Filmstrips

Murray-Darling Basin	b/w		23s.	C.G.
Journey through N.S.W.			42s.	C.G.

Slides

Sydney	4sl.	per set	6s. 6d.	Sime

AUSTRALIA–QUEENSLAND

Agent General for Queensland, 409 Strand, London W.C.2
Answers to questions. Map of Queensland. Book: *Queensland.*
Catalogue of slides and films for free loan.
Recommended films: 'Brisbane–city of sunshine'; 'Channel country'; 'Marvels in miniature'; 'Spoonful of sugar'.

Association of Agriculture

Sugar cane farm study–maps, diagrams and text with photographs, 25s.

AUSTRALIA–TASMANIA

Agent General for Tasmania, 457 Strand, London W.C.2
Printed information, slides and films on free loan.
Small charge for specialist booklets.

Filmstrip

Tasmania	42s.	C.G.

AUSTRALIA–VICTORIA

Agent General for Victoria, Victoria House, Melbourne Place, Strand, London W.C.2
Brochures, maps, catalogue of films, transparencies and filmstrips.
All of high quality and all free loan.

Filmstrip

S. Victoria and Tasmania	b/w			21s.	E.P.

Slides

Melbourne		4sl.	per set	6s.6d.	Sime

AUSTRALIA–WESTERN AUSTRALIA

Government of Western Australia, West Australia House, 115 Strand, London W.C.2

Very helpful. News topic sheet (class set free). Book of basic facts.
Long list of films, filmstrips, slides.
Low-priced filmstrips available for purchase.

Films

Work and transport in N.W. Australia (BBC TV)		20 min.	*h.* 20s.	Foundation
The cattle carters		30 min.	free loan	P.F.B.

Filmstrip

Mediterranean Australia	b/w		23s.	C.G.

AUSTRIA

Austrian Institute, 28 Rutland Gate, London S.W.7

Very good source books.
Particularly good booklet: *Treasure Chest of Europe*
Catalogue of films and slides for free loan.

Austrian Central Statistical Office, Neneburg, Helden Platz, Vienna

Book of statistics; very comprehensive.

Barclays Bank

Economic report (ask for whole set)

Maps–topographic

1:50,000 and 1:25,000 available to order through Stanfords. (Both series incomplete.)

Film

Farmers of Austria		16 min.	*h.* 30s.	Foundation

Filmstrip

Austria	b/w		23s.	C.G.

Slides

Very large selection made on the continent (catalogue) from Sime.

Wallmap

Map		42 x 30 in.	free loan	C.E.W.C.

BELGIUM

Belgian Embassy, 103 Eaton Square, London S.W.1
Excellent source—state needs. Very wide range of good materials.
Belgian National Tourist Office, 66 Haymarket, London S.W.1
Pamphlets and maps on Bruges and Ghent. Map of Amsterdam. Regional pictures.
Booklets: *The coast, Ardennes, Cities* (buildings). Tourist map.
Pamphlet on educational visits.
Holiday booklets full of pictures.
Institut National de la Statistique, 44 rue de Louvain, Brussels
Catalogue of a vast range of statistical publications.

Maps—topographic

1:50,000 and 1:25,000 series. Available from Stanfords.

Film

Belgium		20 min.		*h.* 40s.	Foundation

Filmstrips

Belgium	b/w			15s.	V.I.S.
Belgium				36s.	E.P.

Slides

Brussels		9sl.	per set	13s. 6d.	Star

Large selection made on the continent from Sime.

Wallcharts

Belgium (28 photographs)		13 x 10 in.	free loan	C.E.W.C.

BERMUDA

Bermuda Travel Information Office, Sackville House, 40 Piccadilly, London W.1
Reference map and information brochure for visitors.

BRAZIL

Brazilian Embassy, Information Service, 15 Berkeley St., London W.1
All help free. Fact sheet with map; pamphlet on coffee and Rio de Janeiro.
Map (1:12,000,000). Report on Brasilia.
Guidebook to the Brazilian regions. Booklet: *Statistical glimpse.*
Barclays Bank
Economic report (ask for whole set)

Maps—topographic

1:1,000,000, 1:500,000 and 1:250,000 series (incomplete) available to order through Stanfords.

Films

People of the Highlands		13 min.	*h.* 40s.	Rank
Brazil		14 min.	*h.* 25s.	Gateway
Life in hot rain forests (Amazon Basin)		14 min.	*h.* 25s.	Gateway
The Amazon	b/w	18 min.	*h.* 20s.	Rank

Filmstrips

Brazil: country and people	b/w			15s.	V.I.S.
Brazil: economy	b/w			15s.	V.I.S.
Everyday life in Brazil				36s.	E.P.
Brazil	b/w	33f.		18s.	Unicorn
Amazon lowlands (Everyday life)				36s.	E.P.
South America: the equatorial forest	b/w			25s.	C.G.
Coffee lands of Brazil				44s.	C.G.
N.E. Brazil: natural regions				44s.	C.G.
Brazilian coffee	b/w			20s.	Hulton
Slides					
Brazil		25 sl.	per set	85s.	Grant

BRITISH ISLES–GENERAL

†United Kingdon Atomic Energy Authority, Public Relations Branch, Library and Information Centre, 11 Charles II Street, London S.W.1
List of literature and photographs available. Films on free loan.
Please write with specific enquiries.

The School Government Publishing Co. Ltd., Darby House, 98 Kingston Rd., London S.W.19
Shell guides.

Midland Bank Ltd., Public Relations, Poultry, London E.C.2
Booklet on Europe: *Spotlight on overseas trade*
Research booklets: *Electrical engineering* (for sixth forms)

Institute of Geological Sciences, Exhibition Rd., London S.W.7
Postcards, slides, publications, lectures, demonstrations, films, survey maps, list of helpful firms and reading lists.
First-class for those starting or building a geology department.

Countryside Commission, 1 Cambridge Gate, Regents Park, London N.W.1
List of free publications on each of the National Parks.

The Post Office, Postal Publicity Branch, St. Martins le Grand, London E.C.1
Project kits and pamphlets on postal services.

The Post Office, Telecommunications Headquarters, 138 Cheapside, London E.C.2
Colouring set, long list of books on telephone services, history of the Post Office. Leaflets on use of postal and telephone services. Interested teachers should write stating their need and the age-range of the children concerned. This material may also be obtained from the Telephone Manager covering the area in which the school is located.

Youth Hostels Association, Trevelyan House, 8 St. Stephen's Hill, St. Albans, Herts.
Very good booklets about regional field studies.

National Council for Social Services, 26 Bedford Square, London W.C.1
Social geography—several publications; catalogue available.
Nature Conservancy, Attingham Park, Shrewsbury, Salop.
Able to give advice on the educational use of nature reserves, including nature trails and similar techniques.

Atlases

The Reader's Digest Complete Atlas of the British Isles (Educational edition)	78s.	Longman
First Venture Atlas—British Isles edition	5s.	Philip
Stanford's Geological Atlas of Great Britain	63s.	Stanfords

Films

E.F.V.A. catalogue 3s.

Filmstrips

Map strip	b/w		21s.	E.P.
Home life in England		60f.	55s. 3d.	McGraw

E.F.V.A. catalogue 3s.

Slides

Set of 22 slides on geographical and historical topics in map form		per set	50s.	S.C.
The geography of Great Britain (5 sets)	16sl.	per set	30s.	S.C.

Catalogue available from S.C.
Long list of slide sets from Star Press. Catalogue free.
Very large range from Sime, Woodmansterne or Walton; mainly of a tourist nature but many of value.

Aerial photographs

Aerofilms Ltd., 4 Albemarle St., London W.1
Enquiries can be made by letter or telephone for specific photographs.
Catalogue (extensively illustrated) 15s. Subjects include:

Section 1—Agriculture, horticulture and forestry

1 Land and its cultivation:
Dykes and land drainage, erosion, farm types, fenland, field patterns, irrigation, land reclamation, open fields, lynchets, stone wall country
2 Agricultural buildings
3 Livestock
4 Agricultural machinery
6 Forestry

Section 2—Archaeology

Barrows, camps, crop markings, forts, lynchets

Section 3—Architecture

1 Classified by function, e.g. housing—all types, market places, mills, power stations, railway stations

continued

2 Built-up areas (planned and unplanned), e.g. bombed, cities, estates, industry, new towns, overcrowded areas, ribbon development, slums, slum clearance, towns–history or morphology

Section 4–Communications

1 Sea and river	3 Rail	5 Lighthouses etc.
2 Air	4 Road	6 Lakes, dams, locks, rivers, canals

Section 6–Human and economic geography

Patterns of settlement, land use, industry, communications

Section 7–Physical

1 Landscape features, glacial, rivers and lakes, coastal, erosion and deposition

2 Faults and fractures, rock types and scenery, engineering

Section 9–Industry

Construction, factories, works, fishing, cargoes, mines. Oil and petroleum. Peat. Quarries. Slag heaps. Electricity, gas, sewage and water.

Section 12–Then and now. Examples:

Accrington	1925 and	1961	Bournemouth	1925 and	1961
Bexhill	1931	1958	Caernarvon	1921	1961
Blackburn	1928	1961	Clacton	1938	1961
Brentford	1938	1961	Edgware	1930	1957
Cardigan	1932	1961	Enfield	1924	1957
Coventry	1939	1961	Ilford	1929	1961
Edinburgh	1927	1949	Perivale	1934	1961
Holyhead	1923	1961	Plymouth	1949	1958
Newcastle on Tyne	1924	1961	Rochdale	1926	1961
Redhill	1920	1958	Slough	1925	1950
Ryde	1928	1961			

Wallmaps and charts

Outline–B. Isles (towns, rivers, contours)	40 x 30 in.	12s.	E.P.
Merchant shipping	30 x 40 in.	8s. 6d.	Pictorial
Our fisheries	30 x 40 in.	8s. 6d.	Pictorial
Coasts of Britain (4 charts)	20 x 15 in. per set	10s. 6d.	Pictorial
Physical–G.B. and Ireland	48 x 71 in.	110s.	Philip
Physical–G.B. and Ireland	43 x 33 in.	42s.	Philip
Geology–B. Isles	43 x 33 in.	42s.	Philip
Relief–B. Isles	43 x 33 in.	42s.	Philip
Graphic relief–B. Isles	34 x 42 in.	42s.	Philip
Political–U.K. and Ireland	48 x 71 in.	110s.	Philip
Physical–England and Wales	62 x 70 in.	110s.	Philip
Physical–England and Wales	46 x 36 in.	42s.	Philip
2-sided Physical/Agriculture	48 x 36 in.	85s.	Macmillan
2-sided Industry/Outline	48 x 36 in.	85s.	Macmillan
Physical–B. Isles (Westermann)	72 x 90 in.	185s.	Grant

continued

Black washable 2-sided England & Wales/Scotland, B. Isles/England & Wales, Wales/B. Isles, County/ Region	54 x 42 in.	each 110s.	Grant
Jig-saw maps		10s. 11d., 11s. 7d., 15s. 7d.	E. J. Arnold
Black outline—B. Isles	40 x 30 in.	12s. 6d.	Hulton

MAPS—TOPOGRAPHIC

(A) ORDNANCE SURVEY MAPS

Ordnance Survey, Romsey Rd., Maybush, Southampton SO9 4DH
Catalogue free.

ONE-INCH SEVENTH SERIES
Prices: Coloured: (flat) 7s. Outline: (flat) 7s.

1 Shetland Islands (Yell & Unst)
2 Shetland Islands (North Mainland)
3 Shetland Islands (Lerwick)
4 Shetland Islands (South Mainland)
5 Orkney Islands (North)
6 Orkney Islands (Kirkwall)
7 Pentland Firth
8 Isle of Lewis (Stornoway)
9 Cape Wrath
10 Tongue
11 Thurso
12 Isle of Lewis & North Harris
13 Loch Inver & Loch Assynt
14 Lairg
15 Helmsdale
16 Wick
17 North Uist
18 Harris
19 Gairloch
20 Ullapool
21 Bonar-Bridge
22 Dornoch
23 South Uist
24 North Skye
25 Portree
26 Lochcarron
27 Strathpeffer
28 Inverness
29 Elgin
30 Banff
31 Peterhead
32 Barra
33 Rhum & part of Skye
34 South Skye & Arisaig
35 Loch Arkaig
36 Fort Augustus
37 Kingussie
38 Grantown & Cairngorm
39 Strathdon
40 Aberdeen
41 Braemar
42 Ballater
43 Stonehaven
44 Coll & Tiree
45 Sound of Mull
46 Loch Linnhe
47 Glen Coe
48 Loch Tay
49 Blairgowrie
50 Forfar
51 Iona & Colonsay
52 Loch Awe
53 Loch Lomond
54 Stirling
55 Perth & Alloa
56 St. Andrews & Kirkcaldy
57 Islay
58 Knapdale
59 Firth of Clyde
60 Glasgow
61 Falkirk & Lanark

continued

62 Edinburgh
63 Dunbar
64 Berwick-upon-Tweed
65 Kintyre
66 Isle of Arran
67 Ayr
68 Biggar, Moffat & Sanquhar
69 Selkirk
70 Jedburgh
71 Alnwick
72 Girvan
73 New Galloway
74 Dumfries
75 Dumfries & Gretna
76 Carlisle
77 Hexham
78 Newcastle upon Tyne
79 Stranraer
80 Kirkcudbright
81 Dalbeattie
82 Keswick
83 Penrith
84 Teesdale
85 Durham
86 Redcar & Whitby
87 Isle of Man
88 Barrow-in-Furness
89 Lancaster & Kendal
90 Wensleydale
91 Ripon
92 Pickering
93 Scarborough
94 Preston
95 Blackburn & Burnley
96 Leeds & Bradford
97 York
98 Market Weighton
99 Hull
100 Liverpool
101 Manchester
102 Huddersfield
103 Doncaster
104 Gainsborough
105 Grimsby
106 Anglesey
107 Snowdon
108 Denbigh
109 Chester
110 Stoke-on-Trent
111 Buxton & Matlock
112 Nottingham
113 Lincoln & Grantham
114 Boston & Skegness
115 Pwllheli
116 Dolgellau
117 Bala & Welshpool
118 Shrewsbury
119 Stafford
120 Burton upon Trent
121 Derby & Leicester
122 Melton Mowbray
123 Spalding
124 King's Lynn
125 Fakenham
126 Norwich
127 Aberystwyth
128 Montgomery & Llandrindod Wells
129 Ludlow
130 Kidderminster
131 Birmingham
132 Coventry & Rugby
133 Northampton
134 Huntingdon & Peterborough
135 Cambridge & Ely
136 Bury St. Edmunds
137 Lowestoft
138/151 Fishguard & Pembroke
139 Cardigan
140 Llandovery
141 Brecon
142 Hereford
143 Gloucester & Malvern
144 Cheltenham & Evesham
145 Banbury
146 Buckingham
147 Bedford & Luton
148 Saffron Walden
149 Colchester
150 Ipswich
151 See 138/151
152 Carmarthen & Tenby
153 Swansea
154 Cardiff
155 Bristol & Newport

continued

156 Bristol & Stroud
157 Swindon
158 Oxford & Newbury
159 The Chilterns
160 London N.W.
161 London N.E.
162 Southend-on-Sea
163 Barnstaple
164 Minehead
165 Weston-super-Mare
166 Frome
167 Salisbury
168 Winchester
169 Aldershot
170 London S.W
171 London S.E.
172 Chatham & Maidstone
173 East Kent
174 Bude
175 Okehampton
176 Exeter
177 Taunton & Lyme Regis
178 Dorchester
179 Bournemouth
180 The Solent
181 Chichester
182 Brighton & Worthing
183 Eastbourne
184 Hastings
185 Newquay & Padstow
186 Bodmin & Launceston
187 Plymouth
188 Torbay
189 Land's End
190 Truro & Falmouth

Note: A reprint of the First Edition of the One-Inch Series is published by David & Charles Ltd., South Devon House, Newton Abbot, Devon. Price per sheet 15s.

TWO-AND-A-HALF-INCH FIRST SERIES
Coloured: (flat) 6s. Outline: (flat) 6s.

SELECTED SHEETS

Coastal features:

SS 64 Coastal lands and related drainage
‡SY 67/77 Portland Isle and Chesil Beach
TA 31 Spurn Head
TF 84 Scolt Head Island
TM 45 Orford Ness
TQ 92 Romney Marsh
SH 53 Spits, estuary features
SS 49 Raised beaches–Rhossilli
NJ 92 Dunes
‡SX 85/95 Drowned estuary–Dartmouth
SY 88 Lulworth Cove
TF 74 North coast–Norfolk
TM 44 Orford Ness
TQ 91 Winchelsea
TR 01 Dungeness
SS 43 Erosion surfaces
NH 75 Coastal features

Glacial features:

NX 15 Drumlin field
SE 78 Lake Pickering
SH 63 Snowdonia
‡NC 25/35 North Scotland, highlands, faults, Moine thrusts
SE 64 Terminal moraine
NN 17 Fiord coast–Fort William
SE 89 Newton Dale overflow channels
NX 48 Glacial uplands
NY 75 Overflow channels

Limestone and chalk:

SO 92 Cheltenham–Cotswold scarp Jurassics
SU 13 Chalk–New Sarum

continued

SU 17 Chalk—Vale of Pewsey
SK 17 Limestone—Tideswell
SK 19 Grit moors—Derwent
SD 77 Ingleborough—Carboniferous Limestone
ST 45 Cheddar Gorge
SU 16 Chalk—Marlborough Downs
SU 44 Chalk landscape—near Andover
SK 18 Edale—Millstone Grit Edge

Miscellaneous physical topics:

SX 68 Granite—Dartmoor
TL 88 Thetford Chase
SK 01 Trent terraces
SO 87 Glacial river terraces—Kidderminster
TG 31 Norfolk Broads
TQ 14 Weald—Leith Hill
NT 53 Extinct volcanoes

Human geography topics:

ST 76 Bath—Georgian spa
TG 20 Norwich—expanding historic town
TL 40 Harlow New Town
TQ 28 Willesden, N.W. London
TQ 38 Hackney, N.E. London
SJ 60 Ironbridge gorge—historical
NS 37 Port Glasgow—estuary industry
NZ 41 Thornaby on Tees—industry
SS 99 Rhondda Valley
SK 38 Sheffield—industrial city
SX 99 Exeter—route centre
SJ 89 Central Manchester
NZ 36 Newcastle upon Tyne
SO 51 Monmouth—confluence town
TA 08 Scarborough
TL 30 Lea Valley—recreation area
TQ 27 Wandsworth, S.W. London
TQ 37 Elephant, S.E. London
SJ 39 Bootle and Liverpool—port city
SK 76 Laxton open fields—historical
NZ 24 Durham City—town site
SE 11 Huddersfield—industrial region
SJ 84 Potteries—industrial city
SU 41 Southampton—port city
ST 57 Bristol—port city
‡NT 27/37 Edinburgh—capital city
SP 08 Birmingham—city
ST 33 Bridgwater—bridge town

Note: The sheets marked ‡ above are available only in the new two-and-a-half-inch Second Series, which is replacing the First Series. Most Second Series sheets cover twice the area of the First Series maps, and the England and Wales sheets show public rights of way. Large sheets:—Coloured: flat 8s. Outline: flat 8s. Small sheets:—Coloured: flat 6s. Outline: flat 6s.

OTHER O.S. SERIES AND MISCELLANEOUS MAPS

In addition to the one-inch and two-and-a-half-inch series, the Ordnance Survey publish quarter-inch, six-inch, twenty-five-inch and fifty-inch series, Geological Survey and Soil Survey maps, memoirs and bulletins, and various other maps including those listed below.

Selected 1:625,000 maps (about 10 miles to the inch):

Each map is on two sheets—North and South.
Price 15s. per sheet except where otherwise shown.
Administrative areas map of Great Britain per sheet 7s.

continued

Gravel and associated sands (1959)–South sheet only
Land classification (1939-42)
Land utilisation (1931-39)
Limestone (1949)
Local accessibility (1947-50)
Physical per sheet 7s.
Population change (1951-61)
Population density (1951)
Rainfall: annual average (1916-1950)
Route planning map (revised annually) per sheet 7s.
Topography–North sheet only per sheet 7s.
Solid geology

(B) LAND UTILISATION SURVEY MAPS

Second Land Utilisation Survey of Great Britain
Scale $2\frac{1}{2}$ inches to 1 mile (1:25,000)
Prepared under the direction of Miss A. M. Coleman.
Available from Stanfords. Price per sheet 15s.
Land Utilisation Survey Handbook 3s. Index to sheets 2s.

13 St. Austell
44 Yarmouth (Isle of Wight)
45 Shanklin
50 Exeter
56 Bournemouth
77 Brighton
91 Southampton
110 Winchester (South)
118 Dungeness
138 Folkestone
140 Lynton
158 Dover
160 Cheddar
174 Deal
175 Llantwit Major
176 Barry
187 Croydon
192 Isle of Thanet
193 Bridgend & Porthcawl
194 Cardiff (West)
205 London (South)
206 Bexley
207 Gravesend
210 Port Eynon
211 Port Talbot
212 Ogmore Valley
213 Pontypridd
214 Newport
224 City of London
226 Basildon
227 Southend-on-Sea
230 Llanelly (South)
231 Swansea
232 Rhondda
233 Mountain Ash
235 Chepstow
244 London (North)
249 Milford Haven
252 Llanelly (North)
253 Pontardawe
254 Hirwaun
255 Merthyr Tydfil
263 Thame
264 Chesham
271 Haverfordwest
274 Pontyberem
278 Abergavenny
279 Monmouth
281 Gloucester
287 Harpenden
355 Bedford (South)
388 Presteigne
425 Aberystwyth
442 Thetford
448 Newtown
452 Wolverhampton (South)
454 Nuneaton

continued

461 Methwold
464 Loddon
465 Lowestoft
481 Downham Market
486 Great Yarmouth
487 Cader Idris
499 Crowland Fens
515 Stafford
525 Fakenham
558 Stoke-on-Trent
570 Penygroes
571 Capel Curig
572 Pentrefoelas
573 Ruthin
586 Caernarvon
587 Bethesda
589 Denbigh
620 Llanerchymedd
637 Wallasey
640 Manchester
654 Doncaster
655 Hatfield Moors
666 Hemsworth
667 Thorne
675 Dewsbury
676 Castleford
677 Goole
679 Kingston upon Hull
687 Leeds
688 Sherburn in Elmet
689 Selby
698 Tadcaster
699 Elvington
708 York (West)
709 York (East)
713 Lancaster
723 Dalton-in-Furness
724 Grange-over-Sands
742 Kirkby Moorside
753 Rosedale
757 Ambleside
793 Spennymoor
794 The Hartlepools
801 Durham City
815 Tyneside
820 Whitley Bay
825 Blyth
838 Coldstream
S 138 Dunfermline

BRITISH ISLES—AGRICULTURE

Flour Advisory Bureau Ltd., 21 Arlington St., London S.W.1
Pamphlets, samples, fact sheets, notes—concerned with all stages of bread-making. Filmstrips from E.F.V.A.

National Dairy Council, National Dairy Centre, John Princes St., London W.1
Catalogue of wide range of aids on dairy farming and milk production.

Quaker Oats Ltd., Southall, Middx.
Information on the processing of oats.

Hovis Ltd., 'Wheatgerm', Macclesfield, Cheshire
Booklet: *Looking at wheat* (limited supplies)

Association of Agriculture
Farm studies:

Fruit and hop farm in Kent	20s.	Arable farm in Norfolk	25s.
Upland farm on Welsh Border	20s.	Farm in Yorkshire Dales	25s.
Dairy farm in West Country	20s.	Hill sheep farm in Scotland	25s.

Ministry of Agriculture, Fisheries and Food
Long list of potential sources of information.
Very useful list of photographic sets and wallcharts for loan, including:
British livestock farming; British farm machinery; Rural Britain.
Also news sheets on fishing ports, fishing methods, fishing craft.
Long booklist on fishing, agriculture and statistical sources.

†Herring Industry Board
Booklets: *The story of the herring; The herring book.* Single copies free, then 6d. each.
Filmstrips on catching, processing and distributing, *h* 6s. each. List available.

Booklets

Animal feedstuffs			free	Unilever

Samples

British Wool Cloth Sample Book (14 samples)		5s. 6d.	I.W.S.
Scottish Wool Sample Book (22 samples)		7s. 6d.	I.W.S.
12 wool samples		3s. 6d.	B. Wool

Films

A crop of problems	20 min.	free loan	Potato
A crop for all seasons	20 min.	free loan	Potato

Filmstrips

The forester			36s.	E.P.
Growing a forest	b/w		15s.	Unicorn
Our herring industry	b/w		6s.	Herring
Trawling	b/w	27f.	12s. 6d.	B.I.F.
Drifting	b/w	30f.	12s. 6d.	B.I.F.
Farm family life in England	b/w	35f.	29s. 9d.	McGraw
The story of British wool			21s.	Foundation

Slides

Farming in Britain	12 sl.	per set	30s.	E.P.

Wallcharts

How potatoes are grown	25 x 37 in.		2s. 6d.	Potato
Fruit farm; dairy farm; arable farm; sheep farm	30 x 40 in.	each	6s. 6d.	Pictorial
British wool cloth	19 x 15 in.		6s. 6d.	I.W.S.
The wheat plant			3s. 6d.	E.P.

Picture sets

Sets of 7-19 photographs in a large range of titles		Min. of Ag.
Poster sets on fishing, co-operatives and machines		Min. of Ag.
All on loan.		
8 sets on varieties of the potato	free loan	Potato

BRITISH ISLES—INDUSTRY

Aluminium Federation, Portland House, Stag Place, London S.W.1
Films and aids—list available, Free teaching booklets.
Ford Motor Company, Advertising and Sales Promotion Dept., Cranes Farm Rd., Basildon, Essex
'Facts from Ford'—pack of excellent teaching materials.

British Steel Corporation, Room 80, 33 Grosvenor Place, London S.W.1
Catalogue of charts, fact sheets, films. Very good indeed.
Shipbuilders and Repairers National Association, 21 Grosvenor Place, London S.W.1
The world's best ships (excellent project document)
British shipbuilding (6th-form account)
Shipbuilding in Britain (pamphlet)
Colour film: 'Ships of the world: built in Britain'. Free loan from C.O.I. (C.F.L.)
National Coal Board, Public Relations Dept., Hobart House, Grosvenor Place, London S.W.1
Film catalogue, fact books, coal booklets.
Good booklet: *Visit to a coalmine.*
Charts about coalmines and distribution of coalfields.

Booklets

Oil in the United Kingdom			
The refinery industry		free	P.I.B.

Slides

Heavy industry in Britain	12 slides	per set 30s.	E.P.
Coal mining	notes & 6 slides	15s.	Longman

Wallchart

Fuel and power in Britain (with notes)	10s. 6d.	Pictorial

BRITISH ISLES—LONDON AND SOUTH-EAST ENGLAND

Port of London Authority, P.O. Box 242, Trinity Square, London E.C.3
Catalogue of brochures, wallcharts, films and filmstrips.
Very good source. See Part One, Section B for examples.
†**London Transport, Publicity Officer, 280 Marylebone Rd., London N.W.1**
Maps, timetables, posters, postcards and booklets available for purchase. Full list on request.
Commission for the New Towns, Glen House, Stag Place, Victoria, London S.W.1
Information on Crawley, Hatfield, Hemel Hempstead and Welwyn Garden City.
†**Covent Garden Market Authority, Assistant Superintendent, Covent Garden Market, London W.C.2**
Pamphlets on history and modern functions. One-hour tour.
Supply calendar. (Letters enclosing stamp.)
Corporation of London, Information Office, St. Paul's Churchyard, London E.C.4
Notes on Corporation, places of interest, wallcharts and teachers' notes.
The Post Office Headquarters, St. Martin's le Grand, London E.C.1
Free leaflet: *The Post Office Tower*

Reed Paper Group Ltd., Education Services Dept., Reed House, 82 Piccadilly, London W.1
A project on paper, built around a Medway mill; really excellent. Contains filmstrip, wallchart, teaching notes, record, samples, booklets. Free loan or purchase 62s. 6d. complete, or 25s. without record.

Harlow Development Corporation, 1 Adams House, The High, Harlow, Essex
Welcome to Harlow—pictorial booklet of special interest to visitors, 10d.
Dyelines of the Master Plan 1s. 10d.
Neighbourhood layouts 2s. 10d.
Harlow New Town Master Plan (August 1952) 13s. 6d.
Harlow Sportcentre Brochure 2s. 10d.
Journal of Town and Country Planning, January issues, 8s. 6d. These have many useful articles and statistics.
Living in Harlow—a fully illustrated booklet of the many aspects of Harlow's social life and amenities, 2s. 10d.
History of Harlow, paperback, 16s.
Colour slides 10s. 6d. per set (2 sets available).
All prices include postage.

Borough of Hemel Hempstead, Town Hall, Hemel Hempstead, Herts.
Street map and guidebook.

Lea Valley Water Company, Bishop's Rise, Hatfield, Herts.
Background information of a geographical nature on water supply. Visits welcome—a co-operative source to be used carefully.

Basildon Development Corporation, Gifford House, Basildon, Essex
Report on revised master plan (42s.). Coloured plan (10s.).
Pamphlet on master plan, brochure on history and development, housing, town centre and industrial layout.
Limited supplies but free.

Bracknell Development Corporation, 38 Broadway, Bracknell, Berks.
Very good master-plan booklet. Report on household survey. Town centre plan.

Milton Keynes Development Corporation, Wavedon Tower, Wavedon, Nr. Bletchley, Bucks.

Stock Exchange, P.O. Box 288, 54 Threadneedle St., London E.C.2
Visits of up to 30 by arrangement; booklets explaining function.

Atlas

Atlas of London and the London Region. Maps mainly cover housing, employment, population and traffic. £35 Pergamon

Maps—topographic

See Ordnance Survey and Land Utilisation Survey lists in *British Isles—General* section.

Recommended L.U.S. sheets:

118 Dungeness	192 Isle of Thanet	264 Chesham
158 Dover	244 London (North)	

Films

Growth of London	b/w	23 min.	*h.* 20s.	Rank
London's food markets		7 min.	*h.* 40s.	Rank
Tilbury—container dock		14 min.	free loan	B.T.
Weald of Kent		23 min.	free loan	P.F.B.
Crossroads London		23 min.	free loan	P.F.B.
S.E. England		20 min.	*h.* 10s.	Foundation
All that mighty heart		24 min.	free loan	B.T.
Canterbury	b/w	21 min.	*h.* 20s.	Rank

Filmstrips

Port of London	b/w		20s.	Hulton
London Docks			20s.	E.S.A.
South England—Physical	b/w		15s.	V.I.S.
South England—Economic	b/w		15s.	V.I.S.
London—development and function	b/w		15s.	V.I.S.
Greater London			44s.	C.G.
Medway Valley			45s.	C.G.
Thames Basin	b/w		20s.	Hulton
South East England		27f.	30s.	Rank
London Sewers	b/w		20s.	Hulton
London Markets	b/w		20s.	Hulton
London Airport	b/w		20s.	Hulton
London			30s.	Hulton
Building London's tubes	b/w		free loan or 12s. 6d.	B.T.
A London docker			36s.	E.P.
Everyday life in London			36s.	E.P.
S.E. England			36s.	E.P.
Thames Basin			36s.	E.P.
Mole Gap and Box Hill			36s.	E.P.

(plus essential O.S. map extracts 1s. each)

Slides

Canterbury	8 sl.	per set	14s. 6d.	E.P.
London	18 sl.	per set	30s.	E.P.
Oxford	8 sl.	per set	14s. 6d.	E.P.
Oxford (2 sets)	9 sl.	per set	13s. 6d.	Star
London from the air	9 sl.	per set	13s. 6d.	Star
London streets, City (2 sets)	8 sl.	per set	15s. 6d.	Woodmansterne
London Airport	3 sl.	per set	6s.	Sime
Rocks and scenery in the Weald	notes & 6 sl.	per set	15s.	Longman
Fruit and hop farm in Kent	notes & 6 sl.	per set	15s.	Longman

continued

Crawley–New Town	notes & 6 sl.	per set	15s.	Longman
London	notes & 6 sl.	per set	15s.	Longman
London's New Towns	notes & 16 sl.	per set	30s.	S.C.
Stevenage New Town	notes & 16 sl.	per set	30s.	S.C.

Wallmaps and charts

London (Westermann)	44 x 56 in.	120s.	Grant
Plastic relief (S.E. England)	38 x 26 in.	70s.	O.U.P.
Billingsgate	30 x 40 in.	8s. 6d.	Pictorial

BRITISH ISLES–SOUTH ENGLAND

Southampton Docks–*see British Transport Docks Board in Part One, Section B.*

Maps–topographic

See Ordnance Survey and Land Utilisation Survey lists in *British Isles–General* section.
Recommended L.U.S. sheet: 91 Southampton

Films

Central Southern England	19 min.	*h.* 35s.	Foundation
Southampton Docks	23 min.	free loan	B.T.

Filmstrips

Dorset coast		43 f.	25s.	V.I.S.
Southern England	b/w		20s.	Hulton

Slides

Oil (at Fawley)	notes & 6 sl.	per set	15s.	Longman
Southampton	notes & 6 sl.	per set	15s.	Longman

BRITISH ISLES–SOUTH-WEST ENGLAND

Central Electricity Generating Board, S.W. Region, Public Relations Office, 12-13 Oakfield Grove, Clifton, Bristol 8
Illustrated literature, posters on generation and transmission.
Port of Bristol Authority, Queen Square, Bristol 1
Information about Britain's largest municipal port, comprising City Docks, Avonmouth Dock in Bristol and Portishead Dock in Somerset, and including details of proposed development in 'Bristol Western Docks Scheme'.
Note: Information will only be sent to schools, colleges, libraries, etc., and not to individuals.
Imperial Tobacco Co., Bristol 3
History of the company. Tobacco trade. Overseas growing areas. Also charts and information on tobacco as a topic.
Board of Trade, South West Regional Office, Pithay, Bristol
High-level information on industry and trade.
Limited staff, so please use only if essential.
Dartmoor and Exmoor National Park, Devon County Council, County Hall, Exeter
Large number of picture leaflets–list available.

Exeter Information Office, 18 Queen Street, Exeter
Guides.
Association of Agriculture
Farm study: Dairy farm in West Country, 25s.

Maps—topographic

See Ordnance Survey and Land Utilisation Survey lists in *British Isles—General* section.

Films

South West England	21 min.	*h.*	40s.	Foundation
Berkeley Power Station	38 min.		free loan	A.E.I.
The Cotswolds (BBC TV)	20 min.	*h.*	20s.	Foundation
Severnside pipeline	14 min.		free loan	P.F.B.

Filmstrips

Village study	b/w	39 f.		18s.	Rank
Lower Severn Basin				36s.	E.P.
South-West England				36s.	E.P.
Slapton Ley (plus essential O.S. extract, 1s.)				36s.	E.P.
Scilly Isles				30s.	Unicorn
South West England	b/w			15s.	V.I.S.
South West England		34 f.		30s.	Rank
Cotswolds				44s.	C.G.
Dartmoor				44s.	C.G.
Mixed farming: a Devon farm				42s.	C.G.
Bristol and Cheddar			free loan *or*	20s.	B.T.

Slides

West Country	8 sl.	per set	14s. 6d.	E.P.
The Wye Valley	9 sl.	per set	13s. 6d.	Star
Rocks and scenery in the S.W. (Cornwall)	notes & 6 sl.		15s.	Longman

Wallmap

Plastic relief	38 x 26 in.	70s.	O.U.P.

BRITISH ISLES—WALES

Cwmbran Development Corporation, Victoria St., Cwmbran, Monmouth, Wales
Ministry of Agriculture, Fisheries and Food
Photographic sets in colour.
Steel Company of Wales Public Relations Office, 26 St. James's Square, London S.W.1
Answer questions about industrial activities; will help real need.
John Summers and Sons Ltd., St. Ermins, Caxton St., London S.W.1
Steel literature; 20-minute film about steel products; arranges visits to Shotton Works.
Richard Thomas and Baldwin, RTB House, 151 Gower St., London W.C.1
Sales literature, leaflets and brochures on steel making, wallcharts and films on free loan.

Maps—topographic

See Ordnance Survey and Land Utilisation Survey lists in *British Isles—General* section.

Recommended L.U.S. sheets:

230 Llanelly (South)	232 Rhondda	487 Cader Idris
231 Swansea	233 Mountain Ash	589 Denbigh

Films

South Wales Ports	20 min.	free loan	B.T.
Pembrokeshire	27 min.	free loan	P.F.B.

Filmstrips

North and Central Wales		36s.	E.P.
South Wales		36s.	E.P.
Industrial South Wales		44s.	C.G.
North and Central Wales		44s.	C.G.
Welsh woollen industry	b/w	9s.	I.W.S.

Slides

Cardiff and district	9 sl.	per set	13s. 6d.	Star
Glaciation in Snowdonia	notes & 6 sl.	per set	15s.	Longman
Industrial South Wales	notes & 6 sl.	per set	15s.	Longman

Wallmaps and charts

Black washable 2-sided (Wales/British Isles)	54 x 42 in.		110s.	Grant
Plastic relief (Cardiff and Wales)	19 x 13 in.		15s.	O.U.P.
Outline (Wales)	7 x 5 in.	per 25	3s. 6d.	E. J. Arnold

BRITISH ISLES—MIDLANDS

British Rail, London Midland Region, Euston House, London N.W.1
Posters, details of electrification, Midland services.

Stewarts and Lloyds Ltd., Corby, Northants.
Filmstrips, films, answers to specific questions. Catalogue available.

Corby Development Corporation, Spencer House, Corporation St., Corby, Northants.
General information, questions answered, fact sheet and details of the 1966 Sample Census.

Dawley Development Corporation, Priorslee Hall, Oakengates, Salop.

Stanton and Staveley Ltd., P.O. Box 72, Near Nottingham
Steel works. Film catalogue, booklets on iron and coke, information leaflets.

†John Player and Son, Castle Tobacco Factory, Nottingham
Booklets, films about the factory. Map: World sources of tobacco.

Maps—topographic

See Ordnance Survey and Land Utilisation Survey lists in *British Isles—General* section.

Films

Coventry (BBC TV)		20 min.	*h.* 20s.	Foundation
The Elgro Story (Lincs. farming)	b/w	21 min.	free loan	Midland

Filmstrips

The Midlands			36s.	E.P.
Agricultural East Midlands		26 f.	44s.	C.G.
Industrial West Midlands		28 f.	44s.	C.G.
The Potteries		25 f.	44s.	C.G.
Midlands	b/w		20s.	Hulton
Coal mining: a Midlands colliery			42s.	C.G.

Slides

Stratford-on-Avon	9 sl.	per set	13s. 6d.	Star
Motor vehicle industry	notes & 6 sl.	per set	15s.	Longman
Arable farm in East Anglia	notes & 6 sl.	per set	15s.	Longman

Maps

Plastic relief	38 x 26 in.	70s.	O.U.P.
Birmingham and the Potteries	19 x 13 in.	15s.	O.U.P.

BRITISH ISLES—EAST ANGLIA

Maps—topographic

See Ordnance Survey and Land Utilisation Survey lists in *British Isles—General* section.

Films

East Anglian holiday		20 min.	free loan	B.T.
Farming with a difference (farm study)	b/w	20 min.	free loan	Unilever
Fenland (BBC TV)		20 min.	hire fee	Foundation
East Anglia		21 min.	*h.* 40s.	Foundation
Village in the wheatfields	b/w	20 min.	*h.* 20s.	Foundation
Wheat in East Anglia	b/w	10 min.	*h.* 10s.	Foundation

Filmstrips

Norfolk farming family		36s.	E.P.
Flatford Mill (plus essential O.S. extract, 1s.)		36s.	E.P.
The Fens—farming	b/w	15s.	V.I.S.
The Fens—towns and industry	b/w	15s.	V.I.S.

continued

Eastern England				36s.	E.P.
Eastern England	b/w	42 f.		20s.	Hulton
East Anglia				42s.	C.G.
The Fens		27 f.		44s.	C.G.

BRITISH ISLES–YORKSHIRE

Coghan Bright Steel Ltd., Hunslet Forge, Leeds 10
Help with enquiries–no printed literature.
Hull Docks–*see British Transport Docks Board in Part One, Section B.*

Atlas

East Riding (Elementary level) 7s. 6d.
From A. Brown and Son, Perth Street West, Hull, Yorkshire.

Maps–topographic

See Ordnance Survey and Land Utilisation Survey lists in *British Isles–General* section.

Films

Hull now (1967)		15 min.		free loan	B.T.
Sheffield made		15 min.		free loan	B.S.C.

Filmstrips

Market town–Richmond	b/w	32 f.		18s.	Rank
The Port of Hull	b/w	35 f.		15s.	V.I.S.
The Aire Valley		34 f.		25s.	V.I.S.
Malham Tarn (plus essential O.S. extract, 1s.)				36s.	E.P.
Everyday life of the Yorkshire miner				36s.	E.P.
Scarborough fishing family				36s.	E.P.
Humber ports: Port and trade	b/w	32 f.		12s. 6d.	V.I.S.
Humber ports: Communications	b/w	32 f.		12s. 6d.	V.I.S.
City of York	b/w	57 f.		15s.	B.T.
Yorkshire Dales	b/w	43 f.		15s.	B.T.
Sheffield	b/w	50 f.		17s. 6d.	Cheshire
The steel industry				44s.	C.G.
Immingham and Grimsby docks	b/w			12s. 6d.	B.T.

Slides

Fishing (at Grimsby)		notes & 6 sl.	per set	15s.	Longman
Sheffield–Steel City		notes & 6 sl.	per set	15s.	Longman

Wallmaps

Lancashire and Yorkshire–Physical		48 x 38 in.		44s.	Philip

BRITISH ISLES—PEAK DISTRICT

Maps—topographic

See Ordnance Survey and Land Utilisation Survey lists in *British Isles—General* section.

Filmstrips

Castleton and Edale		32 f.		36s.	E.P.
The Pennines	b/w			20s.	Hulton
Pennines and people	b/w			20s.	Hulton
Walking in the Peak District	b/w			12s. 6d.	B.T.

Slides

The Peak District	9 sl.	per set	13s. 6d.	Star
Rocks and scenery in the Pennines	notes & 6 sl.	per set	15s.	Longman

BRITISH ISLES—LANCASHIRE AND THE ISLE OF MAN

Lancashire Steel Manufacturing Co. Ltd., Irlam, Manchester
Prepared to answer questions about processes and about the geographical aspects of their activities.
Runcorn Development Corporation, Chapel St., Runcorn, Cheshire
Skelmersdale Development Corporation, High St., Skelmersdale, Lancs.
Textile Council, Education and Information Dept., 3 Alberton St., Manchester 3
Booklet on cotton.
Wallcharts: Cotton flow charts (30 x 40 in.), 2s. 6d.; Cotton districts (30 x 20 in.), 2s.; Textile processes (40 x 30 in.), 3s. 6d.
Films include: 'Cotton—Nature's wonder fibre' 18 min. free loan
'Spinning your future' b/w 30 min. free loan
Samples (box) 7s. 6d. Filmstrip: 'Cotton growing', 10s.
Isle of Man Tourist Board, 13 Victoria Street, Douglas, Isle of Man
Official guide, films and slides. Free loan, catalogue available.
Offers information to teachers' questions.

Maps—topographic

See Ordnance Survey and Land Utilisation Survey lists in *British Isles—General* section.

Films

Lancashire		14 min.	*h.* 30s.	Foundation
Men of the Mersey	b/w	12 min.	free loan	Unilever
Merseyside (BBC TV)		20 min.	*h.* 20s.	Foundation
North West England		18 min.	*h.* 35s.	Foundation

Filmstrips

Family life in Liverpool	36s.	E.P.
Lancastria	36s.	E.P.

continued

The Manchester Ship Canal	b/w			15s.	V.I.S.
Liverpool–port and function	b/w			15s.	V.I.S.
North West England	b/w	44 f.		20s.	Hulton
Merseyside–study of a port		26 f.		45s.	C.G.
South Lancashire				44s.	C.G.
Men of rocks	b/w			12s. 6d.	B.T.

Wallmaps and charts

Lancashire & Yorkshire–Physical		48 x 38 in.	44s.	Philip

BRITISH ISLES–LAKE DISTRICT

Maps–topographic

See Ordnance Survey and Land Utilisation Survey lists in *British Isles–General* section.

Films

Climbing an English mountain (Helvellyn)		8 min.	*h.* 25s.	Gateway

Filmstrips

A Lake District sheep farmer			36s.	E.P.
Langdale (plus essential O.S. extract, 1s.)		32 f.	30s.	E.P.
Lake District			36s.	E.P.
English Lakeland	b/w	55 f.	15s.	V.I.S.
Lake District			44s.	C.G.
Lake District and Pennines	b/w		25s.	C.G.

Slides

The Lake District	8 sl.	per set	14s. 6d.	E.P.
The Lake District	9 sl.	per set	13s. 6d.	Star
Sheep farm	notes & 6 sl.	per set	15s.	Longman

BRITISH ISLES–NORTH-EAST ENGLAND

Consett Iron Co. Ltd., Consett, Co. Durham

Economic and sociological studies rather than geographical. Enquiries by letter.

Dorman, Long and Co. Ltd., GPO Box 1, Zetland Road, Middlesborough, Yorkshire

Will provide relevant information regarding activities.

Aycliffe Development Corporation, Churchill House, Beveridge, Newton Aycliffe, Co. Durham

Master plan and advice to advanced students.

Peterlee Development Corporation, Shotton Hall, Castle Eden, Peterlee, Co. Durham
Washington Development Corporation, Usworth Hall, Washington, Co. Durham
North East Development Council, 20 Collingwood St., Newcastle upon Tyne 1
Specific questions rather than general enquiries.
Films on free loan: This is the North East (20 min.); Move North (20 min.); Land of three rivers (20 min.).

Maps–topographic

See Ordnance Survey and Land Utilisation Survey lists in *British Isles–General* section.

Film

Coal and steel in N.E. England		20 min.	*h.* 20s.	Foundation

Filmstrips

North East England			36s.	E.P.
The River Tyne	b/w	30 f.	20s.	Hulton
North East England	b/w	40 f.	20s.	Hulton
Northumberland holiday	b/w	57 f.	15s.	B.T.
Hill lands of northern England			44s.	C.G.
The industrial north east	b/w		free loan *or* 12s. 6d.	B.T.

Wallmap

Plastic relief	38 x 26 in.	70s.	O.U.P.

Slides

The North East	notes & 6 sl.	per set 15s.	Longman

BRITISH ISLES–SCOTLAND

Irvine Development Corporation, Perceton House, Irvine, Ayrshire
Literature and brochures.
Cumbernauld Development Corporation, Cumbernauld House, Cumbernauld, Glasgow
List of publications with prices available. Examples:
planning proposals 21s.; photographs 5s.; air photographs 6s. 6d.; housing survey 10s. 6d.; town plan 1s.
Glenrothes Development Corporation, Woodsides, Glenrothes, Fife
Helpful to genuine enquiries; good books on the town.
Livingstone Development Corporation, Livingstone, West Lothian

Maps–topographic

See Ordnance Survey and Land Utilisation Survey lists in *British Isles–General* section.

Films

Highland estate		21 min.		free loan	Ford
Herrings for sale	b/w	13 min.		free loan	Unilever
Scottish Highlands (BBC TV)		20 min.	*h.*	20s.	Foundation
Border weaver		16 min.	*h.*	40s.	Rank
Glasgow belongs to me		17 min.		free loan	B.T.
Scottish Highlands		20 min.		free loan	B.T.

Very extensive catalogue from Films of Scotland, 3 Randolph Crescent, Edinburgh 3

Examples (all 20 min., hire fee 30s.):

Beyond the Grampians (fishermen, farmers of Aberdeenshire)
Clyde (changing pattern of life)
In good heart (farming year)
Life in the Orkneys
New life (Glenrothes)
Rivers at work (Scottish h.e.p.)

Extensive catalogue from Scottish Film Committee, 16 & 17 Woodside Terrace, Glasgow C.3

Filmstrips

The highland crofter			36s.	E.P.
Central lowlands			36s.	E.P.
Highlands and islands			36s.	E.P.
The Isle of Skye		57 f.	25s.	V.I.S.
Isle of Lewis	b/w	20 f.	20s.	Hulton
Staffa and Iona		53 f.	20s.	B.T.
Scottish highlands	b/w	45 f.	25s.	C.G.
Highland journey		26 f.	40s.	C.G.
The Garth area		35 f.	36s.	E.P.
(plus essential O.S. extract 1s.)				
Life in the Hebrides		25 f.	42s.	C.G.
Southern uplands		36 f.	36s.	E.P.
East Central Scotland	b/w	30 f.	26s.	C.G.
West Central Scotland	b/w	30 f.	26s.	C.G.
Southern Scotland	b/w		25s.	C.G.
British port: Glasgow	b/w	33 f.	18s.	Rank
Edinburgh and Glasgow		26 f.	44s.	C.G.
Wool from the Tweed			42s.	C.G.
Edinburgh			20s.	B.T.
Clyde coast	b/w		12s. 6d.	B.T.

Slides

Very large range from Sime. Each 2s. 6d. All in colour.

Shipbuilding on the Clyde	notes & 6 sl.	per set	15s.	Longman
Scottish highlands	notes & 6 sl.	per set	15s.	Longman

Wallmaps and charts

The highlands of Scotland	30 x 40 in.	free loan	C.O.I.
Scotland	30 x 40 in.	free loan	C.O.I.
Plastic relief	38 x 26 in.	67s. 6d.	O.U.P.
Political	43 x 33 in.	42s.	Philip
Physical	36 x 46 in.	42s.	Philip

BRITISH ISLES—NORTHERN IRELAND

†Government of Northern Ireland, Information Service, Stormont Castle, Belfast

Wallcharts, notes, films.

Tourist Information Office, 6 Royal Avenue, Belfast 1

Irish Linen Guild, Morley House, 314 Regent St., London W.1

See entry in Part One, Section B.

Maps—topographic

1:2,500; 1:10,560; 1:26,720 (N. Ireland and Rep. of Ireland); 1:63,360 (N. Ireland) from Stanfords.

Film

Ireland		20 min.	*h.* 40s.	Foundation

Filmstrips

Irish Linen	b/w	20s.	Hulton
About Ireland		free loan	B.T.
N. Ireland—Physical	b/w	15s.	V.I.S.
N. Ireland—Industry	b/w	15s.	V.I.S.
Flax field to linen fabric	b/w	15s.	V.I.S.

BRITISH ISLES—REPUBLIC OF IRELAND

see Ireland

BURMA

Burmese Embassy, 19a Charles St., London W.1

Free magazine, *Forward*, full of useful facts and figures. Also contains plenty of photographs.

Films

People of the river	b/w	14 min.	*h.* 20s.	Rank
Pattern of Burma		27 min.	free loan	P.F.B.

Filmstrips

Life in Burma		40 f.	55s. 3d.	McGraw
Progress in South-east Asia		36 f.	15s.	Foundation
Burma	b/w		25s.	C.G.
Logging with elephants			25s.	Hulton
The jungle team			*h.* 7s.	M.M.S.

Slides

Series of slide folios	per slide	2s. 6d.	S.C.

CANADA–GENERAL

†Information Services, Canada House, Trafalgar Square, London S.W.1
Catalogue of publications, pamphlets, books of facts, map of Canada, fact sheets, films, photographs on loan–no charge.

Canadian Dairy Foods Service Bureau, 147 Davenport Rd., Toronto 5, Ontario

Canada Department (Agriculture), Information Division, Ottawa, Ontario

Atlases

Canadian Oxford School Atlas	21s.	O.U.P.
Historical Atlas of Canada	18s.	Philip
Senior Canada Atlas	18s.	Philip

Maps–topographic

1:50,000; 1:125,000; 1:250,000; 1:500,000; 1:1,000,000
Also good selection of thematic maps. Available from Stanfords.

Films

Canada's new farmlands	b/w	21 min.	*h.*	20s.	Rank
Canada	b/w	15 min.	*h.*	25s.	Gateway
Wheat supply of Canada	b/w	18 min.	*h.*	20s.	Rank
Timberlands	b/w	16 min.	*h.*	20s.	Rank
Tomorrow's timber (BBC TV)		20 min.	*h.*	20s.	Foundation

Filmstrips

Forestry		42s.	C.G.
Life among the Eskimos		42s.	C.G.
Life on a Prairie farm		42s.	C.G.
Eastern and Northern		36s.	E.P.
Interior and Western		36s.	E.P.
General	b/w	21s.	E.P.
St. Lawrence region	b/w	21s.	E.P.
Prairies	b/w	21s.	E.P.
Shield and North	b/w	21s.	E.P.
Pacific to Atlantic	b/w	15s.	V.I.S.
Great Lakes waterway	b/w	15s.	V.I.S.

Long list produced by the National Film Board of Canada and distributed by S.S. at 33s. each.

Slides

Trans-Canada highway	9 sl.	per set	13s. 6d.	Star

Large selection from Sime (catalogue).
Long list for free loan from C.I.
Long list produced by the National Film Board of Canada and distributed by S.S. at 20s. per set.

Wallmaps and charts

Visual summary	30 x 40 in.	8s. 6d.	Pictorial
Graphic relief	37 x 47 in.	8s.	C.I.
Topography	30 x 41 in.	5s.	C.I.
Canadian forest products	36 x 24 in.	free loan	C.E.W.C.
Canadian timber (12 photographs)	8 x 6 in.	free loan	C.E.W.C.
Black washable 2-sided (Canada/World)	54 x 42 in.	110s.	Grant
ditto (Canada/S. Ontario, Quebec)	54 x 42 in.	110s.	Grant
Physical	73 x 48 in.	110s.	Philip
Political	72 x 48 in.	110s.	Philip
Canada (Westermann)	84 x 78 in.	185s.	Grant
Physical	43 x 33 in.	42s.	Philip
Graphic relief	34 x 42 in.	42s.	Philip

CANADA–ALBERTA

Administrative Officer, Government of Alberta, Alberta House, 37 Hill St., London W.1

Road maps and general pamphlets.

Alberta Department of Lands and Forests, Edmonton, Alberta

North Western Pulp and Paper Ltd., Hinton, Alberta

Department of Industry and Development, 514-11th Avenue, S. W. Calgary, Alberta

Filmstrips

New farmlands (Peace River)	b/w	49 f.		18s.	Rank
Canadian prairies	b/w			22s. 6d.	C.G.

Slides

Alberta	4 sl.	per set	6s. 6d.	Sime

CANADA–BRITISH COLUMBIA

†Agent General for British Columbia, British Columbia House, 1 Regent St., London S.W.1

Free information; films on free loan; descriptive literature on life and industry. Class sets available.

British Columbia Department of Lands and Forests, Victoria, British Columbia

Films

The salmon industry of B. Columbia	b/w	7 min.	*h.*	12s. 6d.	Rank
Canada	b/w	15 min.		25s.	Gateway

Filmstrips

British Columbia	b/w		23s.	C.G.
Western mountain region		42 f.	57s. 6d.	McGraw
British Columbia	b/w		21s.	E.P.

Slides

Vancouver	4 sl.	per set	6s. 6d.	Sime
Vancouver	5 sl.	per set	7s. 6d.	Star

Wallcharts

Train in the Rockies	20 x 30 in.	3s. 9d.	C.I.
Forest region	30 x 40 in.	9s.	C.I.

CANADA–MANITOBA

Hudson's Bay Company, Hudson's Bay House, 79 Main St., Winnipeg, Manitoba
Manitoba Department of Mines and Resources, Winnipeg, Manitoba
Department of Industry and Commerce, Tourist Development Branch, Manitoba
Department of Industry and Commerce, Forest Management Branch, 808 Morquay Building, Winnipeg
Manitoba Department of Agriculture and Conservation, Winnipeg, Manitoba
Association of Agriculture
Farm study (grain farm) 25s.

CANADA–MARITIME PROVINCES

Newfoundland Department of Mines and Resources, St. Johns, Newfoundland
Nova Scotia Department of Lands and Forests, Halifax, Nova Scotia
New Brunswick Department of Lands and Mines, Fredericton, New Brunswick
Prince Edward Island Director of Extension Services, Department of Agriculture, Charlottetown, Prince Edward Island
Association of Agriculture
Farm study (Annapolis Valley of Nova Scotia) 25s.

Filmstrips

The Atlantic Region		42 f.	57s. 6d.	McGraw
Maritime Provinces and Newfoundland	b/w		21s.	E.P.
Newfoundland and Labrador	b/w		26s.	C.G.

CANADA–ONTARIO

Director of Public Relations, Province of Ontario, Ontario House, 13 Charles II St., London S.W.1
Enquiries answered; brochures and information sheets; (no charge for single sheets); official road maps.
Ontario Department, Room 1302, Public Relations and Publicity, East Block, Parliament Buildings, Toronto 2
Ontario Department of Agriculture, Parliament Buildings, Toronto 2
Steel Canada Ltd., 520 Commissioners St., Toronto, Ontario

Canada Copper and Brass Development Association, 55 Yonge St., Toronto 1
Norada Mines Ltd., 44 Kings St., West Toronto 1
Algoma Steel Corporation Ltd., Sault Ste Marie, Ontario
Association of Agriculture
Farm study (mixed farm in South Ontario) 25s.

Filmstrips

Southern Ontario	42 f.		57s. 6d.	McGraw
Lakes Peninsula of Ontario			44s.	C.G.
Great Lakes and St. Lawrence Seaway	b/w		25s.	C.G.

Slides

Ottawa, Niagara Falls, Toronto, Hamilton, Ontario (5 sets)	4 sl.	per set	6s. 6d.	Sime

CANADA–QUEBEC AND LABRADOR

Quebec Department of Agriculture, Quebec, Province of Quebec
Association of Agriculture
Farm study (dairy farm in French Canada) 25s.

Filmstrips

The Canadian Shield	42 f.		57s. 6d.	McGraw
French Canadians	47 f.	*h.*	18s.	Rank
Makkovik: village in Labrador	33 f.		25s.	V.I.S.

Slides

Quebec, Montreal (2 sets)	4 sl.	per set	6s. 6d.	Sime

CANADA–SASKATCHEWAN

Agent General for Saskatchewan, 28 Chester Street, Belgrave Square, London S.W.1
Fact sheet, highways map. Booklet: *The story of wheat.*
Saskatchewan Department of Agriculture, Regina, Saskatchewan

CANADA–YUKON AND THE NORTH

Department of Travel and Publicity, Whitehorse, Yukon, Box 939

Filmstrip

Uranium mining	44s.	C.G.

CARIBBEAN

see West Indies and Caribbean

CEYLON

*Ceylon Tea Centre, 22 Regent St., Piccadilly Circus, London S.W.1
Booklet: *The story of Ceylon tea*
Pamphlets: *Tea from Ceylon,* 6d.; *All in a cup of tea,* 1s. 6d.; *How Ceylon tea is marketed,* 1s. 6d.; *The story of tea,* 1s. 6d.

continued

Tea clippers (2 colour pictures)	15 x 10 in.	6d.
Maps: Ceylon reference map	30 x 20 in.	1s. 3d.
Main areas of tea	20 x 30 in.	1s. 3d.

Project notes: 'How we get our tea'
All very good.
Barclays Bank
Economic report (ask for whole set)

Atlases

Ceylon Atlas (in Sinhalese)	(price on application)	Gunasena
Oxford College Atlas for Ceylon	9s.	O.U.P

Maps

1:63,360 series and thematic maps. Available from Stanfords.

Films

The enchanted island		28 min.	free loan	S.S.
A field in Asia (Ceylon farm)	b/w	28 min.	free loan	U.K.C.F.H.C.

Filmstrips

Ceylon and its people	b/w		17s. 6d.	V.I.S.
Ceylon		30 f.	30s.	Rank
Ceylon	b/w	40 f.	25s.	C.G.
Ceylon tea plantation			42s.	C.G.

Slides

Colombo	4 sl.	per set	6s. 6d.	Sime
Ceylon	12 sl.	per set	41s.	Grant

Wallcharts

Tea	30 x 40 in.	6s.	E.P.
10 mounts	14 x 18 in.	free loan	C.E.W.C.

CHILE

Chilean Embassy, 3 Hamilton Place, Park Lane, London W.1
Free information to specific enquiries.
Barclays Bank
Economic report (ask for whole set)

Films

Fishermen in Quintero	b/w	10 min.	free loan	C.O.I. (C.F.L

Filmstrips

Chile		36s.	E.P.
Chile	b/w	15s.	V.I.S.
Chile		57s. 6d.	McGraw
Chile and Peru	b/w	25s.	C.G.

Maps

List of general and thematic maps from Stanfords.

CHINA

Collet's Chinese Bookshop, Great Russell St., London W.C.1
Books and booklets.
Barclays Bank
Economic report (ask for whole set)

Atlas

China in maps	12s.	Philip

Films

Children of China	b/w	11 min.	*h.* 12s. 6d.	Rank
Children in China		19 min.	*h.* 50s.	Gateway
Village children of South China	b/w	10 min.	*h.* 10s.	National

Very long list of films available from E.T.V., including geography, science, industry, agriculture and cultural topics.

Filmstrips

China 1	b/w		21s.	E.P.
China 2	b/w		21s.	E.P.
Co-operative farming in China			42s.	C.G.
Farmers and boatmen of South China			42s.	C.G.
Journey in China			42s.	C.G.
Oriental City (Canton)		46 f.	18s.	Rank
Hong Kong	b/w		12s. 6d.	V.I.S.
Climate, agriculture	b/w		12s. 6d.	V.I.S.
Farm family life in China	b/w	35 f.	29s. 9d.	McGraw
Ming Li of China		40 f.	55s. 3d.	McGraw
Farming in South China	b/w	47 f.	18s.	Rank

Wallmaps and charts

China and the Orient today	30 x 40 in.	10s. 6d.	Pictorial
China (3 photographic charts)	30 x 20 in.	1s. 9d.	Nelson
China–Physical and communications	46 x 36 in.	44s.	Philip

CLIMATE, METEOROLOGY AND VEGETATION

See also under Meteorological Office Library in Part One, Section B

Films

Life in a Mediterranean climate (California)	b/w	10 min.	*h.* 15s.	Gateway
Life in an oasis (N. Africa)	b/w	10 min.	*h.* 15s.	Gateway
Life in hot rain forests (Amazon Basin)	b/w	14 min.	*h.* 25s.	Gateway

Filmstrips

For Primary children:

All my seasons	A hot day			
A rainy day	A foggy and windy day			
A snowy day	Day and night			
		40 f.	each 57s. 6d.	McGraw
What will the weather be?				
The four seasons		30 f.	each 51s.	McGraw

For Secondary children:

We learn about weather				
Changes in weather				
Understanding weather conditions		50 f.	each 55s. 3d.	McGraw
Clouds and weather		36 f.	30s.	Unicorn
Continental climatic regions	b/w		25s.	C.G.
Mediterranean climatic regions	b/w		25s.	C.G.
Hot deserts	b/w		25s.	C.G.

Long list from E.F.V.A. Catalogue 3s.
Catalogue from Diana Wyllie Ltd., 3 Park Rd., Baker St., London N.W.1

Slides

Cloud formations	12 sl.	per set 41s.	Grant

Overhead projector transparencies

Jan. and July temperatures (6 trans.): Europe; N. America; S. America; Asia; Australasia; Africa	each 24s.	Philip
Evolution of the Monsoon (2 trans.): December to May; June to November	each 24s.	Philip

Wallmaps and charts

Meteorological Record (12 monthly pages on 6 sheets)	$15 \times 17\frac{1}{2}$ in.	9s.	Philip
Annual rainfall–World	44×33 in.	42s.	Philip
Climate maps: World (May-Oct.); World (Nov.-April); Europe; Asia; N. America; S. America; Africa; Australia & N. Zealand	44×33 in.	each 42s.	Philip
Natural vegetation maps: World; Europe; Asia; N. America; S. America; Africa; Australia & N. Zealand	44×33 in.	each 42s.	Philip
Five charts: Precipitation; Formation of clouds; Fronts; Pressure and Winds; Weather map	30×40 in.	per set 52s. 6d. each 12s.	E.P.

continued

Cold lands and seas				
Temperate lands				
Hot lands				
Equatorial forests				
Coniferous forests				
Savannah grasslands				
Temperate grasslands				
Hot deserts	30 x 40 in.	each	8s. 6d.	Pictorial
Mediterranean regions				
The world of water				
Behind the weather map	30 x 40 in.	each	10s. 6d.	Pictorial

Thirty colour posters of cloud photographs, with captions, available from Ford Motor Co. Ltd., Advertising and Sales Promotion Dept., Cranes Farm Rd., Basildon, Essex.

COMMONWEALTH

Commonwealth Institute, Kensington High St., London W.8
Commonwealth Institute, Scotland, 8 Rutland Square, Edinburgh EH1 2AS
Large range of services—very helpful. Send for leaflets.
Royal Commonwealth Society, Northumberland Avenue, London W.C.2
Answer specific questions and offer library to serious students.

Maps—topographic

Large-scale mapping of most Commonwealth countries (excluding the Dominions) and of former British territories prepared by the Directorate of Overseas Surveys and available from Stanfords.

CYPRUS

Ministry of Information, Nicosia, Cyprus
Free information.

Maps

Official survey map	13 x 14 in.	1s. 3d.	C.I.

List of available maps from Stanfords.

CZECHOSLOVAKIA

Czechoslovakia Travel Bureau, 45 Oxford St., London, W.1
Very good selection of leaflets on the regions of the country, often with maps and diagrams. Posters, films, colour slides for free loan.

Filmstrips

Farm family life in Czechoslovakia	b/w	35 f.	29s. 9d.	McGraw
Farms and towns in Slovakia	b/w	45 f.	18s.	Rank

Slides

Karlsbad, Marienbad, Kutna Hora, Brno (4 sets)	4 sl.	per set	6s. 6d.	Sime

Wallcharts

Set of 14	15 x 29 in.	free	C.E.W.C.

DENMARK

†Royal Danish Embassy, 29 Pont St., London S.W.1
Able to reply with printed material to most enquiries.
Booklets include: *Facts about Denmark; Gronlands; Faeroerne; Danish industry in facts and figures; Denmark in pictures; Dansk design.*
Det Danske Selskab (The Danish Institute), 3 Doune Terrace, Edinburgh 3
Ready to assist on cultural matters.
Danish Agricultural Producers, Information Service, Danish Centre, 2-3 Conduit St., London W.1
Films, information sheets, booklets on Danish agriculture.
Only standard replies—no answering questions.
Danish Tourist Board, Sceptre House, 169/173 Regent Street, London W.1
Tourist literature, maps of various parts of the country.
Danish Statistical Department, Fredericksholms, Kanal 27, Copenhagen K
Catalogue of statistical materials.
Nyt Nordisk Forlag, Kobmagerade 49, Copenhagen
Ask for geographical books catalogue (illustrated). Good Danish atlas.
Danish geographical schoolbooks:
Gyldendals Arbejdskort (i) Sjaelland (ii) Fyn og Sydjylland (iii) Nordjylland (iv) Danmark.
Geodaetisk Institut, Copenhagen
National map producers.
1:50,000 series (not yet complete)—very good. Write to Stanfords, stating town or area required.

Filmstrips

Denmark: joined by the sea		27 f.	30s.	Rank
Introduction to Denmark			42s.	C.G.
Denmark	b/w		15s.	V.I.S.
Denmark			36s.	E.P.
Denmark (made in Denmark)	b/w	42 f.	18s.	Unicorn
Denmark		39 f.	30s.	Hulton

Slides

Elisinore, Odense, Copenhagen (3 sets)	10 sl.	per set	15s.	Sime
Copenhagen	9 sl.	per set	13s. 6d.	Star

Wallmaps and charts

Denmark (3 charts)	42 x 30 in.	free	C.E.W.C.
Black washable 2-sided (Alpine countries/Denmark)	54 x 42 in.	110s.	Grant
Physical (Westermann—in German)	82 x 76 in.	185s.	Grant

EGYPT (United Arab Republic)

Films

The Nile in Egypt		13 min.	*h.* 50s.	Gateway
The Suez Canal		17 min.	*h.* 50s.	Gateway
Egypt and the Nile		16 min.	*h.* 40s.	Britannica

Film cassette (8 mm)

Village irrigation		3 min. 35 sec.	69s.	Gateway

Filmstrips

Everyday life in Egypt			36s.	E.P.
The Nile	b/w		15s.	V.I.S.
The Nile		31 f.	39s. 6d.	Gateway
Village life, Town life, Ancient monuments, Suez Canal		31 f.	each 39s. 6d.	Gateway
Family life in Egypt	b/w	35 f.	29s. 9d.	McGraw
Selim of Egypt (Primary level)		40 f.	55s. 3d.	McGraw
Nile Valley	b/w	47 f.	23s.	C.G.
The Nile	b/w	50 f.	15s.	V.I.S.
A journey down the Nile		30 f.	45s.	C.G.
Life along the Nile		32 f.	30s.	Britannica

Slides

Egypt	25 sl.	per set	85s.	Grant
Alexandria, Cairo, Aswan (3 sets)	5 sl.	per set	7s. 6d.	Sime
Alexandria, Cairo, Aswan (3 sets)	9 sl.	per set	13s. 6d.	Star

EUROPE

The British Society for International Understanding, Benjamin Franklin House, 36 Craven St., London W.C.2

Enquiry service on world affairs and individual countries.

European Community Information Service, 23 Chesham St., London S.W.1

Booklet listing current publications. Very good indeed.

The British Atlantic Committee, Benjamin Franklin House, 36 Craven St., London W.C.2

Corporate membership for schools 40s. per annum.

Barclays Bank

Economic Report on the Common Market (ask for whole set).

Booklet

Oil in Western Europe, free from P.I.B.

Atlas

Atlas of Central Europe (Bertelsmann)	45s.	Murray

Films

The changed face of Europe 1945-64		29 min.	*h.* 15s.	C.O.I.(C.F.L.)
No trade barriers	b/w	11 min.	free loan	Dunlop

Filmstrips

Europe	b/w	21s.	E.P.
Europe (12 maps)			B.I.F.
Miners of Europe	b/w	23s.	C.G.
Farmers of Europe	b/w	23s.	C.G.
Fishermen of Europe	b/w	23s.	C.G.
Oxford Projected Atlas		34s.	C.G.

Slides

Geographical slides on every European country made by Westermann and distributed by Grant.

Europe map slides (Primary level)	25 sl.	I.V.A.C.
Europe map slides (Secondary level)	26 sl.	I.V.A.C.
European economy	25 sl.	I.V.A.C.

Wallmaps and charts

Oil for Europe	30 x 40 in.		free	Esso
Geography of Europe (16 charts)	30 x 40 in.	per set	80s.	E.P.
Agriculture	30 x 40 in.		10s. 6d.	Pictorial
Developments since 1945	30 x 40 in.		10s. 6d.	Pictorial
Industrialisation	30 x 40 in.		10s. 6d.	Pictorial
The North Sea	30 x 40 in.		10s. 6d.	Pictorial
Physical/Outline (2-sided)	48 x 36 in.		85s.	Macmillan
Agriculture/Economics (2-sided)	48 x 36 in.		85s.	Macmillan
Population/Communications (2-sided)	48 x 36 in.		85s.	Macmillan
Black washable 2-sided (Europe/World)	54 x 42 in.		110s.	Grant
Political	69 x 62 in.		110s.	Philip
Physical	69 x 62 in.		110s.	Philip
Plastic relief	44 x 30 in.		97s. 6d.	Philip
Central Europe—Physical and communications	46 x 36 in.		44s.	Philip
Outline (plastic)	30 x 40 in.		15s.	E.P.
Black outline	30 x 40 in.		12s. 6d.	Hulton
Wall Atlas (5 maps)—Relief, climate, vegetation, political, population	43 x 33 in.	per set	170s.	
		per map	42s.	Philip
Political (Westermann)	79 x 74 in.		175s.	Grant
Graphic relief	42 x 34 in.		42s.	Philip

continued

Europe	40 x 28 in.	105s.	E. J. Arnold
Pupils' maps	14 x 10 in.	5s. 6d.	E. J. Arnold

FAR EAST

Booklet

Oil in the Far East			P.I.B.

Atlas

Atlas of South-East Asia		50s.	Macmillan

Films

Life in the Far East	16 min.	*h.* 50s.	Gateway

Filmstrips

Everyday life in the East			
Transport in the East			
Everyday life in Malaya and Singapore			
Everyday life in Japan			
Rice growing in Malaya			
Rubber production on a Malayan plantation		each 30s.	E.P.

Wallmaps and charts

Eastern world (2 charts)	30 x 40 in.	per set 15s.	E.P.
Daily Telegraph map of the Far East	30 x 40 in.	3s. 6d.	Geographia
East and South-East Asia (Westermann)	74 x 80 in.	185s.	Grant
Physical and communications	37 x 47 in.	44s.	Philip

FINLAND

†Finnish Embassy, 66 Chester Square, London S.W.1

Large colour pictures (e.g. Helsinki harbour, timber by river, national costume, modern architecture). National map (1:3,000,000). Introductory pamphlets with facts and figures.

Luonnontieteelliinen Seuna, c/o Tieteellisten Seurain Kirjesta, Snellmaninkatu 9-11, Helsinki 17

Finnish Geographical Association for specialist information.

Maps–topographic

1:4,000,000; 1:50,000; 1:20,000. List of available maps from Stanfords.

Films

Lappland	20 min.	free loan	S.S.
A day in a Finnish school	15 min.	free loan	S.S.
Helsinki–trade and industry	15 min.	free loan	S.S.
Helsinki–white city of the north	20 min.	free loan	S.S.
Through ice to Finland	12 min.	free loan	S.S.
Finland's forests converted	25 min.	free loan	T.R.A.D.A.

Slides

Helsinki	4 sl.	per set	6s. 6d.	Sime
Finland	8 sl.			I.V.A.C.

FRANCE

French Railways, French Railways House, 179 Piccadilly, London W.1
Very good booklets and folders on rail travel in various regions of France.
Films on free loan from S.S. (Catalogue available.)
Railway map.
French Government Tourist Office, 178 Piccadilly, London W.1
Free illustrated literature, maps on regions and spas of France.
Set of slides on each of French regions.
French Embassy, Service Culturel, 22 Wilton Crescent, London S.W.1
Film catalogue—long list of geographical films.
'Documentation Pédagogique'—very full catalogue of films, records, slides and brochures. Free loans of up to 1 film, 2 colour slide sets, 3 records (not by post) and 1 filmstrip per loan. Very extensive material on every French region on film and slide.
Institut Français du Royaume-Uni, Queensberry Place, London S.W.7
Extensive and well-documented catalogue available. Fields covered include arts and crafts, geography and travel, Paris and the Paris Region, history, architecture and town planning.
Institut National de la Statistique et des Etudes Economiques, 29 Quai Brandy, Paris 1e
Catalogue of statistical publications.

Books (French sources)

Hachette, 79 Boulevard St. Germain, Paris 6e
Books on France (e.g. La Garonne, Le Rhône, Les Vosges). Regional books (e.g. Provence, Normandy, Brittany).
Géographie 4e: L'Europe et l'URSS (15*f*)—very good French school text-book.
Armand Colin, 103 Boulevard St. Michel, Paris 5e.
Regional books (e.g. Pyrenees, Burgundy, Auvergne, Normandy, Brittany, Corsica).
Hatier, 8 rue d'Assas, Paris 6e.
La France—special book for overseas students.
M.D.I., 29 rue de Fourqueux, St. Germain en Laye, Seine et Oise.

Maps—topographic

1:25,000; 1:50,000; 1:100,000; 1:200,000 (being replaced by new 1:250,000 series). Produced by Institut Géographique National, 107 rue la Boétie, Paris 8e. List of topographical and thematic maps available from Stanfords.

Recommended sheets (1:50,000):

Tours 18/22	Amboise 19/22	Abbeville 22/7
Paris 23/14	Corbeil-Essonnes 23/15	Carcassonne 23/45

continued

Bethune 24/5	Arles 29/43	Crest 30/37
Valreas 30/39	Avignon 30/41	Chateaurenard 30/42
Istres 30/44	Toulon 33/46	
Mt. Blanc 36/31	Castellane 35/42	

Films

Empty seas (Brittany fishing)	b/w	19 min.		free loan	P.F.B.
Rhône Saône Lowland		20 min.	*h.*	40s.	Foundation

Catalogue from French Government Tourist Office (see above) of films for each region and main city. On free loan from S.S.

Film cassette (8 mm)

Paris		4 min.		69s.	Gateway

Filmstrips

Family life in France	b/w	35 f.		29s. 9d.	McGraw
Home life in France		60 f.		55s. 3d.	McGraw
Wine production in France				36s.	E.P.
Everyday life in S.E. France				36s.	E.P.
Paris				36s.	E.P.
Paris		26 f.		30s.	Hulton
Food for the Paris markets	b/w	48 f.	*h.*	18s.	Rank
Pyrenees		26 f.		30s.	Rank
Western Section	b/w			12s. 6d.	V.I.S.
Central Section	b/w			12s. 6d.	V.I.S.
Eastern Section	b/w			12s. 6d.	V.I.S.
North France	b/w	47 f.		25s.	C.G.
Rhône Basin	b/w			25s.	C.G.
S.W. France and Central Massif	b/w			25s.	C.G.
Farming the Paris Basin				42s.	C.G.
Rhône Valley				42s.	C.G.
Journey through S.W. France				42s.	C.G.

Catalogue including filmstrips available from European Schoolbooks.

French sources:

Fernand Nathan, 18 rue Monsieur le Prince, Paris 6e
Armand Colin, 103 Boulevard St. Michel, Paris 5e
Hatier, 8 rue d'Assas, Paris 6e
Marcel Didier, 4-6 rue de la Sorbonne, Paris 5e
E.N.E., 76 rue de Rennes, Paris 6e
Larousse, 17 rue de Montparnasse, Paris 6e

Slides

Box 1. France in four regions	100 sl.	per set	520s.	Harrap
Box 3. Paris	50 sl.	per set	260s.	Harrap
Brittany	5 sl.		7s. 6d.	Sime

continued

Fontainebleau, Paris	2 sets	8 sl.	per set	15s. 6d.	Sime
Paris		9 sl.	per set	13s. 6d.	Star

Catalogue including slides from European Schoolbooks.

French sources:

Publislides, Diapositives Couleurs, Publications filmées d'Art et d'Histoire, 44 rue du Dragon, Paris 6e

Orly (No. 40–6 sl.); Paris (Nos. 35, 36, 37, 38–6 sl. per set)

Hachette, 73 Boulevard St. Michel, Paris 5e

E.N.E., 76 rue de Rennes, Paris 6e

Very large catalogue of slides covering all regions.

Wallmaps and charts

Physical, profiles, climate (2-sided, in French)	48 x 36 in.	85s.	Macmillan
Black washable 2-sided (France/Baltic Sea countries)	54 x 42 in.	110s.	Grant
France (Westermann)	84 x 81 in.	185s.	Grant
Paris (Westermann)	44 x 56 in.	120s.	Grant
France and Low Countries–Physical and communications	36 x 46 in.	44s.	Philip

Catalogue including wall pictures, posters and maps from European Schoolbooks.

French sources:

Cartes Industrielles, 9 rue Coetlogon, Paris 6e

Catalogue of economic maps of France.

Armand Colin, 103 Boulevard St. Michel, Paris 5e

Delagrave, 15 rue Soufflot, Paris 5e

Girard et Barrère, 25bis, rue Henri Barbuse, Paris 5e

Jeulin, 4-8 rue Chauvelot, Paris 15e

André Lesot, 10 rue de l'Eperon, Paris 6e

M.D.I., 29 rue de Fourqueux, St. Germain en Laye, Seine et Oise

Hatier, 8 rue d'Assas, Paris 6e

Series of charts on all French regions and systematic topics for France as a whole:

100 x 120 cm Paper 12 *f.* *Carton* (mounted on cardboard) 40 *f.*
Plastique (mounted on plastic) 55 *f.*

Hachette, 79 Boulevard St. Germain, Paris 6e

Series on French regions, 2-sided 100 x 120 cm

Paper 11 *f.* *Carton* 35 *f.*

Rossignol (Editions), Route de la Trimouille, Montmorillon (Vienne), Paris 16e

Superb series of air photographs in colour, 56 x 76 cm

Wallmaps. Full colour catalogue available.

GAMBIA

Gambia High Commission, Gambia House, 28 Kensington Court, London W.8

Limited requests for brochures.

Commonwealth Institute		
Official survey map	8 x 14 in.	2s.
Photographic set (8 plates)	12 x 15 in.	12s.
Wallmap	36 x 25 in.	12s.

Maps–topographic

1:50,000 and 1:125,000 from Stanfords.

GERMANY

German Embassy, Curzon Public Relations Ltd., 31 St. James's Place, London S.W.1

Excellent materials for all ages on most aspects of Germany. State your needs. Film library catalogue–very good.

Atlases

Diercke World Atlas (Westermann)	52s. 3d.
Regional editions of Westermann atlases:	
Berlin Schools Atlas	12s.
Hamburg Schools Atlas Part 1	12s.
Part 2	20s.
Dortmund Schools Atlas	16s. 6d.
Niedersachsen	24s.
Hesse	26s. 6d.
N. Rhine/Westphalia	26s. 6d.
Schleswig-Holstein	26s. 6d.
Bavaria	11s.
Rhineland-Pfalz	15s. 6d.

All the above have German texts. Available from Grant.

Maps–topographic

1:25,000; 1:50,000; 1:200,000 from Stanfords.

Good selection (1:50,000):

Hamburg L2524	Uchte L3518
Braunschweig L3728	Duisburg L4506
Essen L4508	Koln L5106
Bonn L5308	Koblenz L5710
Freiburg im Breisgau L8112	Dusseldorf L4706
Wuppertal L4708	Treuchtlingen L7130
Solingen L4908	Garmisch Partenkirchen L8532

Films

People of the Industrial West		16 min.	*h.* 40s.	Rank
Children of Germany	b/w	14 min.	*h.* 20s.	Rank
Background to Berlin			*h.* 15s.	C.F.L.
Bavarian village		17 min.	*h.* 50s.	Gateway
The River Rhine		22 min.	*h.* 20s.	Foundation
Full ahead Rhine		16 min.	*h.* 40s.	Rank

Long list from E.T.V. on the Democratic Republic, including geography, science, industry and agriculture. Catalogue available.

Filmstrips

West Germany	b/w	40 f.	18s.	Unicorn
Northern Germany			36s.	E.P.
East Germany	b/w	40 f.	18s.	Unicorn
Central and Southern Germany			36s.	E.P.
History and North Plain	b/w		15s.	V.I.S.
Scenery and people of S. Germany	b/w		15s.	V.I.S.
The Rhine	b/w		15s.	V.I.S.
Farm family life in Germany	b/w	35 f.	29s. 9d.	McGraw
North German Lowlands	b/w	37 f.	23s.	C.G.
Uplands of Central and S. Germany	b/w	39 f.	23s.	C.G.
Western Germany	b/w	35 f.	20s.	Hulton
The Rhine	b/w		25s.	C.G.

Slides

Germany—maps (German or English text)	20 sl.	
Industries and towns (Fed. Rep. Ger.)	25 sl.	
Southern regions (F.R.G.)	27 sl.	
Rhine (F.R.G.)	21 sl.	
N. Germany (F.R.G.)	32 sl.	
Central regions (F.R.G.)	30 sl.	
Berlin E. & W.	16 sl.	
West & S. West	25 sl.	
German Democratic Republic	27 sl.	All from I.V.A.C.

Very large selection of Westermann slides available from Grant. Set on each region. (Catalogue.)

Wallmaps and charts

Federal Republic (3 charts)	25 x 37 in.	per set	9s.	E.P.
The Rhine	30 x 40 in.		10s. 6d.	Pictorial
Democratic Republic (3 charts)	30 x 20 in.		free loan	C.E.W.C.
Physical, political, economic (2-sided, in German)	48 x 36 in.		85s.	Macmillan

Westermann maps (in German), available from Grant:

Germany & neighbours (Political)	185s.	S.W. Germany	185s.
Germany (Physical)	185s.	S. Germany (Bavaria)	185s.
Germany (Economic)	185s.	Rhineland	185s.
Schleswig-Holstein	185s.	Saar	175s.
N. Rhine Westphalia	175s.	Hesse	185s.
Lower Saxony	175s.	Hamburg	320s.
N. Germany	185s.	Berlin	120s.
Black washable (Germany/Europe)	110s.		

SOURCES IN GERMANY

These are really excellent for those requiring up-to-date information. At present local sources and national agencies are very co-operative. Please try to be very exact about the materials required. Write to London offices (address in London telephone directory) or to the Embassy. The Embassy will usually give the address in Germany.

Paul List Verlag KG, 8 Munchen 15, Goethestrasse 43
Regional Atlas of Germany, 15Dm.
City Atlas of German cities, 10Dm.
Wallcharts—all German regions.
Catalogue in English.
Harms World Geography—Deutschland, 35Dm.

Verlag Ferdinand Hirt, 23 Kiel 1, Postfach 280, Schauenburgersstrasse
Deutschland, Heft iii, 6Dm.
Seydlitz Erdkunderwerk, Band 1, Deutschland, 12Dm.

Walter De Gruyter & Co., 1 Berlin 30, Genthinerstrasse 13

Velhagen & Klasing u Herm Schroedal Gmbh & Co. K.G., 1 Berlin 31, Bingerstrasse 62
Ask for catalogue.

Statistisches Bundesant, P.O. Box 828, Wiesbaden
National statistical source.

Dokumentations und Auskuntsdents der Bundesregerung, Bonn, Nassestrasse 11
Regional statistical source.

GHANA

Barclays Bank
Economic report (ask for whole set).

Films

Surf of Accra		15 min.	free loan	Unilever
Ghana today (1964)	b/w	15 min.	*h.* 20s.	Rank

Filmstrips

The Ghana cocoa farmer			36s.	E.P.
Country, people, economy	b/w	18 f.	15s.	V.I.S.
Cocoa and chocolate		35 f.	15s.	Unicorn
Ghana		28 f.	42s.	C.G.
Ghana		30 f.	25s.	Hulton

Maps—topographic

1:50,000; 1:62,500; 1:125,000; 1:250,000 from Stanfords.

Wallmaps and charts

Ghana (2 charts)	37 x 25 in. per set	6s.	E.P.
Cocoa plantation	30 x 40 in.	6s.	E.P.
Cocoa farming	36 x 30 in.	free loan	C.E.W.C.

continued

Black washable 2-sided			
(Ghana/Africa)	54 x 42 in.	110s.	Grant
Physical & communications	34 x 22 in.	44s.	Philip
Portfolio of Ghana maps		40s.	Stanfords

GIBRALTAR

Gibraltar Tourist Office, 15 Grand Buildings, Trafalgar Square, London W.C.2

Tourist information of a good standard.

Films				
Gateway to the Mediterranean	12 min.	*h.*	40s.	Foundation
Filmstrip				
Everyday life in Gibraltar			36s.	E.P.
Slides				
Gibraltar	5 sl.	per set	7s. 6d.	Sime
Maps and wallcharts				
Photographic poster	20 x 30 in.		2s.	C.O.I.
Photographic poster	20 x 30 in.		2s. 3d.	C.I.
Official survey map	7 x 5 in.		1s. 3d.	C.I.

GREECE

Royal Greek Embassy, Press and Information Office, 49 Upper Brook St., London W.1

List of free information sheets, including:

6. Athens–ancient and new city
23. Report on industry (1964-65)
24. Agriculture (1964-65)
26. Greece and the Common Market
27. Mines, quarries and mineral production
29. Greek shipyards
30. Greek economy
35. Statistical data

National Tourist Organisation of Greece, 195/197 Regent St., London W.1

Large range of pamphlets including maps. Please state your needs.

Filmstrips					
Everyday life in Greece				36s.	E.P.
Mediterranean shores	b/w	45 f.		30s.	Rank
Modern Greece	b/w	33 f.		15s.	Foundation
Isles of Greece	b/w	38 f.		15s.	Foundation
Greece		45 f.		30s.	Hulton
Greece	b/w	37 f.		18s.	Unicorn
Slides					
Athens, Delphi (2 sets)		10 sl.	per set	15s.	Sime

Also a large selection made on continent from Sime.

Wallmaps and charts

Danube Basin & Balkan Peninsula–Physical (Westermann–in German)	72 x 90 in.	185s.	Grant

HOLLAND

see Netherlands

HONG KONG

Atlas

Modern School Atlas–Hong Kong edition		16s.	U.P.S.
Supplements only	per ten	20s.	Philip

Maps–topographic

Ministry of Defence 1:10,000; 1:25,000; 1:100,000 from Stanfords.

Films

Hong Kong–free port	b/w	14 min.	*h.* 50s.	Gateway
Crossroads of the Orient		16 min.	*h.* 40s.	Rank
The boat people		14 min.	free loan	P.F.B.

Film cassette (8 mm)

Hong Kong	3 min. 25 sec.	69s.	Gateway

Filmstrips

Hong Kong	b/w	36 f.	15s.	Hulton
Hong Kong		27 f.	30s.	Rank

Slides

Hong Kong	4 sl.	per set	6s. 6d.	Sime
General views (1963-4)			free loan	C.I.

Wallchart

Hong Kong	12 panels	12 x 14 in.	free loan	C.E.W.C.

HUNGARY

British Hungarian Friendship Society, 84a Claverton St., London S.W.1
Free loans of exhibitions, films, reading materials and cultural publications. Loan of books to answer specific questions (postage appreciated). Very helpful.

Films

Children of Hungary	b/w	10 min.	*h.* 10s.	Foundation

Long list of films from E.T.V., including geography, travel, science, industry, agriculture and cultural subjects.

Filmstrip

Hungary	b/w	36 f.	18s.	Unicorn

Slides

Large selection (made by a continental firm) from Sime.

ICELAND

Icelandic Embassy, 1 Eaton Terrace, London S.W.1
Pamphlets, brochures. Films (catalogue)—free loan.
Several good films on vulcanism.
Iceland Tourist Information Bureau, 161 Piccadilly, London W.1
Slides, films, brochures on geology and geography of Iceland.
Barclays Bank
Economic report (ask for whole set).

Maps—topographic

1:50,000; 1:100,000; 1:250,000. Available from Stanfords.

Film cassette (8 mm)

Thermal areas	3 min. 30 sec.		60s.	Macmillan

Filmstrips

General survey		36s.	E.P.
Fishing industry		36s.	E.P.
Iceland	34 f.	30s.	Rank
Iceland	26 f.	30s.	Hulton
Iceland—physical aspects	36 f.	25s.	V.I.S.

Slides

Iceland (4 sets)	4 sl.	per set	6s. 6d.	Sime

Wallcharts

Iceland (2 charts)	20 x 25 in.	per set	5s.	E.P.

INDIA

Indian Information Service, India House, Aldwych, London W.C.2
Annual review; long catalogue of films (heavily booked); picture booklet.
Church Missionary Society
Exhibitions, photographic sets, news panel, study outlines, films, filmstrips and slides.
Methodist Missionary Society
Long list of project kits and films.
Commonwealth Institute
Vast list of materials for free loan. Very up-to-date.
Brooke Bond Tea Ltd., Education Dept., Heathrow House, Bath Rd., Cranford, Hounslow, Middx.
Booklet: *The story of tea*, 1 copy free, then 1s. per copy.
Wallchart (40 x 30 in.) 1s. 6d. Filmstrip (50f.) 10s.

Films

India		22 min.	*h.*	50s.	Gateway
The peaceful revolution		26 min.		free loan	A.E.I.
Blend of India (about tea)		10 min.		free loan	S.S.
Two leaves and a bud		19 min.		free loan	S.S.
Food for thought	b/w	20 min.	*h.*	30s.	Contemporar

continued

India	b/w	17 min.	*h.* 20s.	Rank
Deccan village		10 min.	*h.* 25s.	Rank
Farmers of India	b/w	21 min.	*h.* 20s.	Rank
The rice growers		10 min.	*h.* 20s.	Foundation
Tropical harvest		17 min.	*h.* 30s.	Boulton
Indian minorities	b/w		free loan	India
Indian rubber	b/w		free loan	C.O.I. (C.F.L.)

Film cassettes (8 mm)

Outline topography	55s.	Street scenes	55s.
A market	55s.	Village life	69s.
Village farming	55s.	Hindu religion	55s.

Varying length. All from Gateway.

Filmstrips

Tea growing in India			36s.	E.P.
Everyday life in Indian village			36s.	E.P.
Land and people	b/w		21s.	E.P.
Climate and agriculture	b/w		21s.	E.P.
Forests, fisheries, minerals	b/w		21s.	E.P.
Plain of the Ganges		27 f.	30s.	Rank
History	b/w		15s.	V.I.S.
The land	b/w		15s.	V.I.S.
The people	b/w		15s.	V.I.S.
Deccan village		24 f.	30s.	Rank
The people		40 f.	42s.	Gateway
Village life		42 f.	42s.	Gateway
Town life		45 f.	42s.	Gateway
Ramesh of India		40 f.	55s. 3d.	McGraw
Geography of India; Social institutions; Village India; Urban India		all 43 f.	each 71s. 3d.	McGraw
Village life in India			42s.	C.G.
The Indus Basin	b/w		25s.	C.G.
The Ganges Basin	b/w	45 f.	25s.	C.G.
Peninsular India	b/w	46 f.	26s.	C.G.
India: a changing country		37 f.	*h.* 10s.	M.M.S.
Indian village	b/w	36 f.	20s.	Hulton
Village life	b/w	46 f.	25s.	Foundation
City life in India	b/w	37 f.	25s.	Foundation
Indian city	b/w	31 f.	20s.	Hulton

Slides

Bombay, Calcutta, Delhi, New Delhi (4 sets)	4 sl.	per set	6s. 6d.	Sime
Northern India	25 sl.	per set	85s.	Grant

continued

Central and Southern India	25 sl.	per set	85s.	Grant
India: Rural life	30 sl.			I.V.A.C.
India: Towns, men, religions	30 sl.			I.V.A.C.
India: Kashmir	25 sl.			I.V.A.C.

Wallmaps and charts

India (10 panels)	14 x 18 in.		free	C.E.W.C.
Indian cotton mills (25 panels)	14 x 18 in.		free	C.E.W.C.
Bullock cart	20 x 30 in.		3s. 9d.	C.I.
People and problems	10 x 8 in.			Oxfam
Indian scenes	20 x 30 in.			Nelson
Indian agriculture	30 x 40 in.		10s. 6d.	Pictorial
Rice	30 x 40 in.		10s. 6d.	Pictorial
Middle East and India (Westermann)	81 x 74 in.		185s.	Grant
Topographic	40 x 30 in.		5s. 6d.	C.I.
Outline	7 x 5 in.	per 25	3s. 6d.	E. J. Arnold

INDONESIA

Barclays Bank

Economic report (ask for whole set)

Films

Indonesia today	28 min.		*h.* 20s.	B. & F.

Filmstrips

Life in Indonesia	40 f.		55s. 3d.	McGraw

Slides

Java, Bali, Biak (3 sets)	4 sl.	per set	6s. 6d.	Sime
Indonesia: Java and Bali	20 sl.			I.V.A.C.

IRAN

Barclays Bank

Economic report (ask for whole set)

Filmstrips

Part 1 Outline of history	34 f.		30s.	Rank
Part 2 Economy and industry	34 f.		30s.	Rank

IRAQ

Film

Oil of Iraq	27 min.		free loan	P.F.B.

Filmstrip

Iraq	30 f.		30s.	Rank

IRELAND–NORTHERN

see British Isles–Northern Ireland

IRELAND–REPUBLIC OF

Borde Failte Eireann (Irish Tourist Board), P.O. Box 273, Baggot St. Bridge, Dublin
Map: Ireland 12 miles to 1 inch
Holiday pack for visitors.
Ireland Welcomes–magazine (6 issues per annum) 1s. 6d.
Many local guides on application.

Films

Ireland		20 min.	*h.* 40s.	Foundation

Maps–topographic

1:2,500; 1:10,560; 1:26,720 (N. Ireland and Rep. of Ireland) from Stanfords.

Filmstrips

Part 2 S. Ireland	b/w	25s.	C.G.
About Ireland		free loan	B.T.

ISRAEL

Jewish National Fund, Rex House, 4-12 Regent St., London S.W.1
Large collection of information sheets, e.g. law, boundaries, population, geography, transport and education.

Maps–topographic

1:100,000. In English except for southern Negev sheets (Hebrew). From Stanfords.

Films

Land of promise	b/w	20 min.	*h.* 20s.	Foundation

Filmstrips

Israel	b/w		21s.	E.P.
Building a nation	b/w	42 f.	18s.	Rank
Israel		34 f.	30s.	Hulton

Slides

Jerusalem	9 sl.	per set	15s.	Star

Large selection from Sime.

Wallmaps and charts

Israel	25 x 37 in.	2s. 6d.	E.P.
Palestine–Physical	37 x 46 in.	42s.	Philip

ITALY

Italian State Tourist Office, 201 Regent St., London W.1
Booklets full of information about travel in Italy. Excellent for projects.
Barclays Bank
Economic report (ask for whole set).
†Instituto Italiano di Cultura, 39 Belgrave Square, London S.W.1
Information service with films, records and slides, books and photographs for loan.

Instituto Centrale de Statistica, Via Cesare Balbo 16, Rome
Catalogue of wide range of statistical materials.
UTET, Corso Raffaello 28, Turin
Catalogue of whole series of regional books of Italy. Well illustrated.

Maps—topographic

1:100,000. Also 1:50,000 (new series in progress) and 1:200,000.
Available from Stanfords. Good selection (1:100,000):

Bologna 87	Varese, Lake Maggiore 31
Venezia 51	Firenza 106
Napoli 183-184	

Films

Po river valley	b/w	20 min.	*h.*	20s.	Rank
Apulian village	b/w	10 min.	*h.*	20s.	Foundation
Tuscan countryside	b/w	12 min.	*h.*	20s.	Foundation
People of Venice		15 min.	*h.*	50s.	Gateway

Filmstrips

Everyday life in Italy			36s.	E.P.
Northern Italy			36s.	E.P.
Central and Southern Italy			36s.	E.P.
Farm family life in Italy	b/w	35 f.	29s. 9d.	McGraw
Home life in Italy		60 f.	55s. 3d.	McGraw
Rome	b/w	33 f.	20s.	Unicorn
Po valley	b/w	48 f.	18s.	Rank
Village life in N. Italy		26 f.	42s.	C.G.
Italy	b/w	40 f.	25s.	C.G.
Italian village	b/w	16 f.	20s.	Hulton
Italy	b/w	40 f.	18s.	Unicorn

Slides

Rome, Florence, Venice (3 sets)	9 sl.	per set	15s.	Star
Large catalogue from Sime.				
Sets of cities	10 sl.	per set	30s.	European
The North	32 sl.			
The Centre I	26 sl.			
The Centre II	3 sl.			
The South I	33 sl.			
The South II	5 sl.			
Sicily	20 sl.			
Rome	12 sl.			All from I.V.A.C.

Wallmaps and charts

Physical (Westermann—in German)	63 x 77 in.	185s.	Grant

JAPAN

Japan Information Centre, Embassy of Japan, 9 Grosvenor Square, London W.1
Free information in reply to specific requests.

Barclays Bank
Economic report (ask for whole set).

Films

Japan	13 min.	*h.* 40s.	Rank
Harvesting land and sea	16 min.	*h.* 40s.	Rank
Miracle of Asia	16 min.	*h.* 40s.	Rank
Agriculture in Japan	16 min.	*h.* 40s.	Foundation
Fisheries in Japan	16 min.	*h.* 40s.	Foundation
Changing face of Japan	16 min.	*h.* 40s.	Foundation
Food for Japan	11 min.	*h.* 30s.	Foundation

Filmstrips

Farm family life in Japan	b/w	35 f.	29s. 9d.	McGraw
Country life in Japan		40 f.	55s. 3d.	McGraw
Physical background		27 f.	30s.	Rank
Industry, the ports and people		26 f.	30s.	Rank
A Japanese family	b/w	27 f.	20s.	Hulton
An island nation	b/w	45 f.	18s.	Rank
Japan	b/w	35 f.	20s.	Hulton
Agriculture in Japan		24 f.	21s.	Foundation
Fisheries in Japan		18 f	21s.	Foundation

Wallmaps and charts

Physical/communications	47 x 37 in.	44s.	Philip

Slides

Tokyo, Kamakura, Kobe (4 sets)	4 sl.	per set	6s. 6d.	Sime
Northern Japan	25 sl.	per set	85s.	Grant
Southern Japan	25 sl.	per set	85s.	Grant
General study	25 sl.			I.V.A.C.
Landscapes, agriculture, vulcanism	26 sl.			I.V.A.C.
Populations	30 sl.			I.V.A.C.
Towns, industry, fishing	26 sl.			I.V.A.C.

JORDAN

Slides

Jordan-Syria	25 sl.	I.V.A.C.

Catalogue of sets from Sime.

KENYA

Kenya High Commission, 45 Portland Place, London W.1
Excellent duplicated sheets on the 'Land and People' showing the evolution of the country and the people.
Colour films for free loan include:
Kenya becomes a nation 30 min.
Somali fishing (Kenya rivers) 25 min.

continued

Permanent way (East African railways) 45 min.
Kenya Arabica (African coffee farmer) 30 min.
Booklet: *African peoples of Kenya.* Magazine: *Kenya Today.*
Church Missionary Society
Photographs and news panels available for loan.

Atlas

The First Kenya Atlas	(price on application)		J.K.F.

Maps–topographic

1:50,000 and 1:250,000 series from Stanfords.

Films

Kenya today	b/w	18 min.	*h.* 20s.	Rank

Filmstrips

Coffee in Kenya		33 f.	21s.	Unicorn
Kenya	b/w	35 f.	17s. 6d.	Foundation
Kenya		43 f.	30s.	Hulton

Slides

Coffee in Kenya	free loan	C.I.
General geography	free loan	C.I.
Nairobi	free loan	C.I.
Peoples of Kenya	free loan	C.I.
Tea growing	free loan	C.I.
Mombasa (1962)	free loan	C.I.

Wallmaps and charts

Physical and communications		36 x 46 in.	44s.	Philip
Kenya (26 mounts)	b/w	14 x 18 in.	free loan	C.E.W.C.
Photographic poster		30 x 40 in.	2s.	C.I.
Relief and communications		38 x 34 in.	11s. 10d.	C.I.
Official survey map		15 x 14 in.	1s. 3d.	C.I.

KOREA (SOUTH)

Korean Embassy, 36 Cadogan Square, London S.W.1
Information sheets, copies of Korean news magazine. Good.

Filmstrip

Peasant farming in Korea	34 f.	30s.	Rank

KUWAIT

Films

Close-up on Kuwait	28 min.	free loan	P.F.B.
Story of a town	22 min.	free loan	P.F.B.

LIBYA

Films

Desert drive		15 min.	free loan	Ford
Libya	b/w	15 min.	*h.* 20s.	Foundation

Filmstrip

Oases in Libya	31 f.		30s.	Britannica

Slides

Tripoli, Benghazi, Syrte, Tobruk (4 sets)	4 sl.	per set	6s. 6d.	Sime
Westermann set of geographical slides	12 sl.	per set	41s.	Grant

MALAWI

High Commission for Malawi, 47 Great Cumberland Place, London W.1

Excellent duplicated booklet includes details of housing, health, education, shipping, recreation, churches, etc.—all information for potential visitors.

Barclays Bank

Economic report (ask for whole set).

Maps

1:50,000 series (Directorate of Overseas Surveys). Available from Stanfords.

Official survey map	14 x 8 in.	1s. 3d.	C.I.

Filmstrip

Malawi (1966)	b/w	free loan	C.I.

MALAYSIA

Barclays Bank

Economic report (ask for whole set).

Atlases

Malaysian Modern School Atlas		15s.	U.P.S.
Supplements only	per 10	24s.	U.P.S.
Senior Atlas for Malaysia and Singapore (price on application)			U.P.S.

Maps—topographic

1:25,000 and 1:2,500 (Directorate of Overseas Surveys). Available from Stanfords.

Films

Land of tin and rubber	13 min.	*h.*	40s.	Rank
Jungle pioneer (BBC TV)	20 min.	*h.*	20s.	Foundation
The vital link	18 min.		free loan	P.I.B.
Asian crescent	23 min.		free loan	Unilever

Filmstrips

Rice growing in Malaya			36s.	E.P.
Rubber on a Malayan plantation			36s.	E.P.
Malayan rubber plantation			42s.	C.G.
Set of 8 strips, including physical geography, rubber, padi, coconut, mining, industry, E. coast		each	30s.	Hulton
Rubber	b/w		20s.	Hulton

continued

Malayan village	b/w			20s.	Hulton
Country and people	b/w			15s.	V.I.S.
People at work	b/w			15s.	V.I.S.
Nomads of the jungle	b/w	53 f.		18s.	Rank
Malaya	b/w	44 f.		20s.	Hulton
Everyday life in Malaya and Singapore		37 f.		36s.	E.P.

Slides

Slides and news panel			free loan	C.M.S.
Singapore	4 sl.	per set	6s. 6d.	Sime

Long list for free loan from C.I.

Wallmaps and charts

Malaysia and Singapore—Land use	37 x 47 in.	44s.	Philip
Malaya—Physical and communications	37 x 47 in.	44s.	Philip
Malaya	47 x 37 in.	12s.	C.I.
Malaya—visual summary	30 x 40 in.	8s. 6d.	Pictorial

MALTA

Malta High Commission, Malta House, 26 Haymarket, London S.W.1
Free literature—tourist emphasis. Answer queries. Reference library available.
Director of Education, Valletta, Malta
Information available.

Atlas

Clarion Atlas of Malta	5s.	Longman

Maps—topographic

1:25,000 and 1:2,500 (Directorate of Overseas Surveys) from Stanfords.

Films

Malta	16 min.	*h.*	40s.	Rank

Filmstrips

Malta: Mediterranean miniature	29 f.	30s.	Rank
Malta	13 f.	20s.	Hulton

Slides

Malta	10 sl.	per set	15s.	Sime
Valletta, Gozo (2 sets)	3 sl.	per set	6s.	Woodmanste

Maps and charts

Photographic poster	30 x 40 in.	2s. 3d.	C.I.
Official survey map	5 x 7 in.	1s. 3d.	C.I.

MATHEMATICAL GEOGRAPHY

Film cassettes (8 mm)

Time	Summer and winter			
Longitude	Contours			
Latitude	Night and day		each 55s. to 69s.	Gateway

Latitude
Longitude
Seasons (March)
Seasons (June)
Seasons (Sept.)
Seasons (Dec.)
Causes of wind belts
Determining position
Movement of Earth round Sun
Swing of the pressure systems and wind belts
Day and night
Wind direction
Pressure and winds: monsoon lands
Map projections–graticule
Map projections–conical
Map projections–cylindrical

Map projections–zenithal	All 1 to 3 min.	each 50s.	Rank

Filmstrips

Maps–what they are for
Types and uses
Symbols and terms
Globes
Latitude & longitude–places & directions

Time zones and climate			each 55s. 3d.	McGraw
Map making (age level 8-10):				
The village				
The town				
The big city	b/w	40 f.	each 18s.	Unicorn

Overhead projector transparencies

Latitude and longitude (set of 7)	Rank

Wallchart

Map projections	30 x 40 in.	10s.	E.P.

MEXICO

Barclays Bank
Economic report (ask for whole set).

Films

El Barrio	b/w	15 min.	*h.* 20s.	Contemporary
Mexican children	b/w	11 min.	*h.* 12s. 6d.	Rank

continued

Land and people		16 min.	*h.*	40s.	Rank
Harvest of discovery		10 min.	*h.*	25s.	Rank
Adobe village	b/w	19 min.	*h.*	20s.	Rank
Volcanoes of Mexico		8 min.	*h.*	25s.	Rank

Filmstrips

Mexico			36s.	E.P.
Farm family life in Mexico	b/w	35 f.	29s. 9d.	McGraw
Adobe village	b/w	52 f.	40s.	Rank
Introduction to Mexico		25 f.	42s.	C.G.

MIDDLE EAST

Atlas

Oxford Regional Economic Atlas of the Middle East and N. Africa	70s.	O.U.P.

Booklet

Oil in the Middle East	free	P.I.B.

Film

Middle East	14 min.	*h.*	20s.	Rank

Film cassette (8 mm)

Suez Canal	4 min.	69s.	Gateway

Filmstrips

Middle East	b/w	53 f.	25s.	Foundation
Middle East	b/w	43 f.	25s.	C.G.

Slides

Israel, Jordan, Lebanon (3 sets)	4 sl.	per set	6s. 6d.	Sime
Lebanon	20 sl.			I.V.A.C.

Wallmaps and charts

Near East and Middle East–Physical and communications	37 x 48 in.	44s.	Philip
Daily Telegraph map	30 x 40 in.	5s.	Geographia
Middle East	30 x 40 in.	10s. 6d.	Pictorial
Middle East (11 mounts)		free loan	C.E.W.C.

Long list from V.C.O.A.D.

MOROCCO

Maps–topographic

1:500,000; 1:200,000; 1:50,000 available from Stanfords.

Filmstrips

Desert nomads	b/w	35 f.	18s.	Rank

Slides

Casablanca, Tetuan, Rabat, Tangier, Marrakesh (5 sets)	4 sl.	per set	6s. 6d.	Sime
Morocco	24 sl.			I.V.A.C.

NETHERLANDS

Royal Netherlands Embassy, Dept. of Information, 38 Hyde Park Gate, London S.W.7

Answers specific enquiries. Superb map of Netherlands. Factual booklets on Zuider Zee works, history of shipping, etc.

Picture folders on The Hague, Rotterdam, Amsterdam.

Barclays Bank

Economic report (ask for whole set).

Books and booklets

Portrait of Holland	12s. 6d.	E.P.
Carriers of Europe	8s. 6d.	E.P.
Grendel von Holland	17s. 6d.	

from Co-op Nederland, Rotterdam 7, Vierhavens Straat 40

Maps—topographic

Topografische Dienst, Delft, Westvest 9

Catalogue (very good) and maps available from Stanfords.

Recommended sheets (1:50,000):

37 Rotterdam (2 sheets)	25 Amsterdam (2 sheets)
21 Zwolle (2 sheets)	7 Groningen (2 sheets)
48 Middelburg (2 sheets)	40 Arnhem (2 sheets)

Films

Children of Holland	b/w	11 min.	*h.* 12s. 6d.	Rank
Gateway to Europe		15 min.	*h.* 40s.	Rank
Land out of Sea	b/w	20 min.	*h.* 20s.	Foundation
Hold back the Sea		30 min.	free loan	P.F.B.

Filmstrips

Everyday life in the Netherlands			36s.	E.P.
The Netherlands			36s.	E.P.
Home life in Holland		60 f.	55s. 3d.	McGraw
The Netherlands	b/w		15s.	V.I.S.
Gateway to Europe		32 f.	30s.	Rank
The Netherlands	b/w	44 f.	23s.	C.G.
Land behind the dykes	b/w	50 f.	18s.	Rank
Holland	b/w	40 f.	18s.	Unicorn
Reclamation of the Zuider Zee		28 f.	42s.	C.G.

Dutch filmstrip producers:

Fibo-color Filmstrips, Utrechtseweg 66-68, Zeist

Ask for catalogue.

Nederlandse Projectie Centrale, Multicolor Slides, 9 Great Chapel St., London W.1

N.V. Filmfabrick Polygoon, Postbox 508, Hilversum

Slides

Amsterdam, Rotterdam (2 sets)	9 sl.	per set	13s. 6d.	Star
Landscapes I	30 sl.			I.V.A.C.
Landscapes II	5 sl.			I.V.A.C.
Dwellings	25 sl.			I.V.A.C.
Industry, transport, fishing	24 sl.			I.V.A.C.

Catalogue of 98 sets, covering every large Dutch town, from Film Bureau Niestadt, Burgem, Gulcherlaan 36, Hilversum.

Wallmaps and charts

Netherlands, Belgium & Luxembourg (Westermann)	54 x 88 in.	185s.	Grant
Netherlands	20 x 30 in.	9s.	E.P.

NEW ZEALAND

New Zealand High Commission, New Zealand House, Haymarket, London S.W.1

Film library (catalogue). Reference library (loan service).
Free loan (up to 3 weeks) of class sets of 10-30 books (one set per loan) for age-levels 6-8, 8-11, and Secondary. Large selection (catalogue).
Fact sheets: geography, history, farming, Maori people, flora, fauna.
Good information booklets, including New Zealand Dairy Industry.
Picture sheets also available.

Association of Agriculture

Dairy farm study—maps, diagrams and text with photographs, 25s.

Australia and New Zealand Bank Ltd., 71 Cornhill, London E.C.3

Book: *New Zealand prospects.*

New Zealand Dairy Board, St. Olaf House, Tooley St., London S.E.1

Booklet: *Dairy farm in New Zealand.*
Details of manufacture of dairy produce (rather technical).

New Zealand Information Bureau, Williams House, Eastbourne Terrace, London W.2

Booklet: *History of New Zealand lamb.*

Barclays Bank

Economic report (ask for whole set).

Atlas

Oxford Atlas for New Zealand	22s. 6d.	O.U.P.

Maps—topographic

1:63,360 (incomplete); 1:25,000 (incomplete).
List of New Zealand maps available from Stanfords.

Films

Life on a N.Z. sheep farm		14 min.	*h.* 50s.	Gateway
N. Island dairy farm		14 min.	*h.* 50s.	Gateway
New Zealand today	b/w	15 min.	*h.* 20s.	Rank

Filmstrips

N. Island				36s.	E.P.
S. Island				36s.	E.P.
Natural background	b/w			21s.	E.P.
Human aspects	b/w			21s.	E.P.
Country and climate	b/w			15s.	V.I.S.
People and work:					
Sheep and cattle	b/w			15s.	V.I.S.
Industry and transport	b/w			15s.	V.I.S.
New Zealand (2 parts)	b/w	40 f.	each	25s.	C.G.
Life in New Zealand		28 f.		42s.	C.G.
A dairy farm in New Zealand (2 parts)	b/w	38 f.	each	15s.	V.I.S.
Land of the Maoris	b/w	38 f.		15s.	V.I.S.

Slides

New Zealand	12 sl.	per set	41s.	Grant
N. Island	9 sl.	per set	13s. 6d.	Star
S. Island	9 sl.	per set	13s. 6d.	Star

Wallmaps and charts

Apple growing	25 x 37½ in.		2s. 6d.	E.P.
New Zealand (10 mounts)	14 x 18 in.		free loan	C.E.W.C.
New Zealand (25 photographs)	12 x 15 in.		free loan	C.E.W.C.
New Zealand (6 charts)	30 x 20 in.		free loan	N.Z.H.C.
Outline	7 x 5 in.	per 25	3s. 9d.	E. J. Arnold
Physical	43 x 33 in.		42s.	Philip
Political	43 x 33 in.		42s.	Philip
N. Island–Physical	43 x 33 in.		42s.	Philip
S. Island–Physical	43 x 33 in.		42s.	Philip

NIGERIA

National Westminster Bank
Economic report.
Church Missionary Society
Exhibition, photographic sets, pictorial news, films, slides.

Maps—topographic, etc.

Stanfords have a very large selection, including:
Set of 25 maps scale 1:3,000,000 of the nation under headings Agriculture, Communications, etc., each 10s.
1:50,000 Topographic series
1:12,500 Town plans: Calabar, Enugu, Ilorin, Jos, Kano, Onitsha, Port Harcourt, Zaria
1:30,000 Lagos
Series of small-scale maps of the regions of Nigeria.

Films

Project Nigeria (rubber development)		15 min.	free loan	Dunlop
Traders in leather		14 min.	free loan	Unilever
West Africa (Nigeria)		18 min.	*h.* 40s.	Rank
The twilight forest		27 min.	free loan	Unilever
The oil rivers		17 min.	free loan	Unilever
People and produce	b/w	10 min.	*h.* 10s.	Foundation
Search for oil		26 min	free loan	P.F.B.
Framework for a nation		31 min.	free loan	P.F.B.
Enterprise in Nigeria		21 min.	free loan	Unilever
Toward the highlife		22 min.	free loan	Unilever

Filmstrips

Contrasts in Nigeria			free loan	C.I.
East and West Nigeria			free loan	C.I.
Everyday life in Nigeria			36s.	E.P.
Northern Nigeria			36s.	E.P.
East and West Nigeria			36s.	E.P.
Nigeria	b/w		15s.	V.I.S.
Contrasts in Nigeria		36 f.	30s.	Britannica
Northern Nigeria		38 f.	30s.	Hulton
Village life in N. Nigeria		28 f.	42s.	C.G.

Slides

Journey across Nigeria (1965)		
Lagos (1964)		
West Nigerian life (1965)		
General views (1964)		
North Nigeria (1964)		
Products (1966)	All free loan	C.I.

Wallmaps and charts

Official survey map	15 x 14 in.	1s. 3d.	C.I.
Wallmap	37 x 48 in.	12s.	C.I.
Diorama	30 x 18 x 11 in.	1s. 6d.	C.I.
Black washable 2-sided (Nigeria/West Africa)	54 x 42 in.	110s.	Grant
Physical and communications	42 x 33 in.	44s.	Philip
Giant in the sun	18 x 31 in.	2s.	S. Times
Nigerian industry (17 mounts)	12 x 15 in.	free loan	C.E.W.C.

NORTH AMERICA

see Canada and United States

NORWAY

Royal Norwegian Embassy, Cultural Attaché, 25 Belgrave Square, London S.W.1

Filmstrips and colour slides on loan. Filmstrips also for sale.

Books on most topics concerned with Norway may be borrowed.
Single copies of booklets and tourist maps available.
Free loan of classroom sets of pictures.

Central Bureau of Statistics (Statistisk Sentralhyra), Drenningengate 16, Oslo

Statistical source—very helpful if needs are specified and information required is contained within one of their publications.

National Westminster Bank

Economic report

Dreyers Forlag, Arbiensgate 7, Oslo

Norwegian geography books in English for Primary schools. Catalogue available.

Norges Geografiske Oppmaling, St. Olavgate 32, Oslo

National map producers. Catalogue available—especially fine are the 1966 *et seq.* 1:50,000 topographic series.
Topographic maps are also available from Stanfords.
There are a number of sheets available of a new series in four or five colours, very superior to the black and white edition.
Glacial and geological maps also available.
Ask for list or state cover needed.

Films

Norwegian children	b/w	11 min.	*h.*	12s. 6d.	Rank
Farmer fishermen	b/w	20 min.	*h.*	20s.	Rank
Southern Norway	b/w	16 min.	*h.*	15s.	Foundation
Return to Norway		25 min.		free loan	P.F.B.

Filmstrips

Norway				36s.	E.P.
Everyday life in W. Norway				36s.	E.P.
Farmer fishermen	b/w	50 f.		18s.	Rank
Life in Lappland		33 f.		30s.	Rank
Norway		30 f.		30s.	Rank
Norway	b/w			17s. 6d.	V.I.S.
Elle Kari, Lapp girl	b/w	32 f.		18s.	Unicorn
Norway	b/w	50 f.		23s.	C.G.
Norway	b/w	37 f.		17s. 6d.	Foundation
Fiords and mountains	b/w	30 f.		18s.	Unicorn
Norway (2 parts—Danish producer)	b/w	40 f.	each	18s.	Unicorn
Tree to paper	b/w			20s.	Hulton
Train across the mountains	b/w	48 f.		20s.	Hulton
Geography of Norway	b/w	49 f.		20s.	Hulton
Norwegian fiord		26 f.		42s.	C.G.

Slides

Norway: extreme north of Europe	14 sl.	I.V.A.C.

Wallmaps and charts

Set of 10 mounts	15 x 22 in.	free loan	C.E.W.C.
Scandinavia & Baltic Lands–Physical	36 x 46 in.	44s.	Philip

PAKISTAN

National Westminster Bank
Economic report.

Atlas

Oxford School Atlas for Pakistan			7s	O.U.P.

Film

Oil in Pakistan	19 min.		free loan	P.F.B.

Filmstrips

East Pakistan			36s.	E.P.
West Pakistan 1			36s.	E.P.
West Pakistan 2			36s.	E.P.
Farm family life in Pakistan	35 f.		29s. 9d.	McGraw
Pakistan: East and West	45 f.		30s.	Britannica
Irrigation in W. Pakistan	28 f.		42s.	C.G.

Slides

Karachi	4 sl.	per set	6s. 6d.	Sime
Pakistan	12 sl.	per set	41s.	Grant
Pakistan: Kashmir	25 sl.			I.V.A.C.

Long list for free loan from C.I.

Wallcharts

Pakistan (10 mounts)			free loan	C.E.W.C.
Pakistan (4 diagrams)			free loan	C.E.W.C.

PANAMA

Film

The Panama Canal	b/w	11 min.	*h.*	15s.	Gateway
Panama Canal: gateway to the World		14 min.	*h.*	40s.	Rank

Filmstrips

The Panama Canal	b/w	29 f.		18s.	Unicorn

PERU

Maps–topographic, etc.

1:200,000 and 1:100,000 series
Geological map (2 sheets)
Available from Stanfords.

Films

Arid zones of Peru		22 min.		hire fee	C.O.I. (C.F.L.)
Forgotten Indians	b/w	31 min.		hire fee	C.O.I. (C.F.L.)
Peru	b/w	11 min.	*h.*	12s. 6d.	Rank

Filmstrips

Peru	b/w			15s.	V.I.S.
Market day at Cusco		40 f.		55s. 3d.	McGraw
Peru		43 f.		57s. 6d.	McGraw
Highlands of the Andes	b/w	56 f.		18s.	Rank
Peru	b/w	41 f.		18s.	Unicorn
Pedro drives the llamas		12 f.		20s.	Hulton

Slides

Lima		4 sl.	per set	6s. 6d.	Sime

PHILIPPINE ISLANDS

Films

The Philippines	b/w	10 min.	*h.*	10s.	Foundation

Filmstrips

Life in the Philippine Islands		40 f.		55s. 3d.	McGraw
Philippine Islands	b/w	32 f.		20s.	Hulton

PHYSICAL GEOGRAPHY

Institute of Geological Sciences, Exhibition Rd., London S.W.7
Postcards, slides, lectures, demonstrations, films, survey maps, reading lists, list of helpful firms. First-class for those starting or developing a physical geography or geology department. Recent additions include overhead projector transparencies.

Marion Ray, 36 Villiers Avenue, Surbiton, Surrey
Twenty sheets of fossil drawings, per set 40s. Filmstrip: Fossils, 20s. Five filmstrips on soils, 20s. each: Rocks and erosion; Plants and decay; Soil profiles; Composition of soil; Soil and farming.

Films

The ice watchers	18 min.		free loan	S.S.
Case history of a volcano	30 min.	*h.*	63s.	Gateway
The beach	20 min.			
Evidence for the Ice Age	19 min.			
Levelling by erosion and deposition	14 min.			
Rocks that form on the earth's surface	16 min.			
Rocks that originate underground	22 min.			
Waves on the water	16 min.			
Against the sea	17 min.	All *h.*	40s.	Rank

Film cassettes (8 mm)

Caverns	Sedimentation			
Lakes	Volcanoes			
A dyke	Lava flows			
Oceans	Moraines			
Potholes	Tectonic movements			
Mountains		All $3\frac{1}{2}$ to $4\frac{1}{2}$ min.	each 95s.	Macmillan

Filmstrips

Glaciation	Influence of rocks			
Vulcanicity	Coastal development			
Granite country	Limestone & chalklands			
Agents of change	Mountains–growth & decay			
Soil erosion	Oceans & islands			
Rivers and valleys				
	b/w		each 17s. 6d.	V.I.S.
Rocks	The earth's surface			
Land & sea	Mountain building			
Denudation	Transport & deposition			
Rivers & lakes	Volcanoes & earth movements			
Useful minerals				
	b/w		each 12s. 6d.	V.I.S.
Work of the sea	Landforms & coastal movements			
Work of rivers	Natural vegetation of Britain			
Desert landforms	Glaciation in Highland Britain			
Volcanoes	Weathering & erosion			
Volcanic features			each 36s.	E.P.
Glaciers (no notes)	Geological action of wind and ice			
Earthquakes	Geological action of running water			
Geology of sea and lakes				
	b/w		each 21s.	E.P.
Rivers	Mountains & hills			
Snow & ice	Oceans, seas, lakes			
The seaside	Valleys & plains			
	b/w		each 23s.	C.G.
Lakes	Landscape of normal erosion			
Volcanoes	Landscape of glacial erosion			
The sea coast	River development			
Desert landforms	Weathering & mass wasting			
Mountain building	Glaciers & ice sheets			
Limestone scenery				
	b/w		each 25s.	C.G.
Africa: Rift valley, volcano, gorge			each 39s.	C.G.
N. America: Glacier, waterfall, canyon			each 39s.	C.G.
Historical geology of B. Isles			44s.	C.G.
Work of the sea	Volcanoes			
Work of ice	Weathering			

continued

Work of rivers	Earth's crust			
	b/w		each 20s.	Hulton
Igneous rocks			30s.	Hulton
Exploring our earth	High & low lands			
Story of a river	Lakes & rivers			
		44 f.	each 55s. 3d.	McGraw
Coastlines 1—physical		29 f.		
2—human response		25 f.		
Glaciation		29 f.		
Vulcanicity		24 f.	each 30s.	Rank
Glaciers	b/w	22 f.	18s.	Rank

Slides

Philips' Physical Geography Colour Slides—99 sets of 6 slides each, with notes. List available. per set 15s.
per slide 3s. 6d. S.C.

Rocks, landforms and scenery	Chalk scenery			
Glacial features	Limestone scenery			
Coastal deposition				
	16 sl.	per set	30s.	S.C.
The aspects of land relief	13 sl.			
Water: its motion and effects	18 sl.			
Types of coastal formation	14 sl.			
Virgin forest to eternal snow	17 sl.			
Glaciers	18 sl.			
Glacial relief	25 sl.			
Volcanic activity	24 sl.			
Volcanic relief: technical uses	18 sl.			
Limestone relief and karst phenomena	26 sl.			
Fluvial erosion	36 sl.			
Erosion of the soil	20 sl.			
Seashore morphology: accumulation	12 sl.			
Seashore morphology: erosion	30 sl.			All from I.V.A.C.

Wallmaps and charts

Carboniferous limestone scenery
Chalk scenery
Deserts
Mountain building
Glaciation
River valleys
Sea coasts
Science and a river
The world of the soil
The world of water
Lands beneath the waves

continued

Catastrophes			
Rivers and man	All with notes each	10s. 6d.	Pictorial
The moon		8s. 6d.	Pictorial
Stone and wall		8s. 6d.	Pictorial
600 million years of earth history		free	Esso

POLAND

Polish Cultural Institute, 16 Devonshire St., London W.1
Free publications (send for list) giving a good picture of Polish life and ideas today.

Maps–topographic, etc.

Provincial sheets 1:500,000 (17 sheets).
Physical map. Geological map. Motoring map.
Road atlas.
Available from Stanfords.

Film

Poland	b/w	22 min.	*h.* 20s.	Rank

Filmstrip

Poland	b/w	41 f.	18s.	Unicorn

POLAR

Films

Summer at the South Pole (1967)		10 min.	free loan	P.F.B.
Antarctic crossing (1958)		49 min.	free loan	P.F.B.

Filmstrips

Life in the polar regions		44 f.	55s. 3d.	McGraw
The Antarctic	b/w	42 f.	21s.	E.P.
Arctic village	b/w	19 f.	20s.	Hulton
Commonwealth trans-Antarctic Expedition			36s.	E.P.

Film cassettes (8 mm)

Arctic ice flows	4 min. approx.	69s.	Gateway
Eskimo village	4 min. approx.	69s.	Gateway
Travel in the Arctic	4 min. approx.	69s.	Gateway

Maps and wallcharts

List of maps for Antarctica and Falkland Islands and Dependencies available from Stanfords.

Falkland Islands Official Map	12 x 19 in.	1s. 3d.	C.I.
Grahamland in summer		8d.	C.I.
The South Atlantic (3 charts)	20 x 25 in.	5s.	E.P.
The Far North (10 mounts)	14 x 18 in.	free	C.E.W.C.

PORTUGAL

Maps—topographic, etc.

Topographic series in progress at scales 1:100,000 and 1:50,000.
List of geological and other maps available from Stanfords.

Filmstrips

Portugal			36s.	E.P.
Portugal		38 f.	30s.	Rank
Portugal	b/w	58 f.	15s.	V.I.S.
Port wine from Portugal	b/w	29 f.	20s.	Hulton
Vineyard in Portugal		12 f.	20s.	Hulton
Portugal	b/w	40 f.	18s.	Unicorn

Wallcharts

Portugal (4 charts)	25 x 37 in.	per set 12s.	E.P.
Portugal and her overseas territories	30 x 40 in.	8s. 6d.	Pictorial

RUSSIA

see U.S.S.R.

SAUDI ARABIA

Filmstrips

Ali of Saudi Arabia (Primary level)	40 f.	55s. 3d.	McGraw

Maps—topographic

General and communications maps (small scale).
1:100,000 series (Directorate of Overseas Surveys).
Available from Stanfords.

SINGAPORE

Wallmaps and charts

Singapore (6 plates)	15 x 12 in.	8s. 4d.	H.M.S.O.
Malaysia and Singapore—Land use	37 x 47 in.	44s.	Philip

SOUTH AFRICA

Information Attaché, South African Embassy, Trafalgar Square, London W.C.2

Booklets and information sheets. Please state subject: General survey, Education, Separate development, Fact sheet, Deserts.

Deciduous Fruit Board, Publicity Officer, Norman House, 105 Strand, London W.C.2

Booklet: *The South African deciduous fruit industry.*

South African Foundation, 7 Buckingham Gate, London S.W.1

Aims to promote understanding of South Africa and her problems. Topics covered include: education, climate, vegetation, transport.

Cape Fruit, Southern Postal and Duplicating Services, 63 Old Town, Clapham, London S.W.4
Booklet: *The story of Cape fruit.*
Films on Cape fruit on free loan from S.S.

Atlases

Atlas of the Republic of South Africa	approx.	£30 from	Stanfords
Venture Atlas for Southern Africa		8s.	Philip
Junior School Atlas for Southern Africa		4s.	Philip
Large-print Atlas for Southern Africa		11s.	Philip
Students' Atlas for Southern Africa		14s.	Philip
College Atlas for Southern Africa		16s.	Philip
Supplements only	per ten	30s.	Philip

Maps—topographic

Series at scales from 1:1,000,000 to 1:50,000. All around 15s.
Available from Stanfords.

Films

Mining Centre—Johannesburg	b/w	16 min.	*h.*	20s.	Rank
The Union of South Africa		16 min.	*h.*	40s.	Rank
Riches of the Veld	b/w	19 min.	*h.*	20s.	Rank

Filmstrips

Everyday life in South Africa				36s.	E.P.
Native peoples of South Africa				36s.	E.P.
Citrus growing in South Africa				36s.	E.P.
South Africa (2 parts)			each	36s.	E.P.
Country, climate and people	b/w			15s.	V.I.S.
Economy and farming	b/w			15s.	V.I.S.
Minerals and industry	b/w			15s.	V.I.S.
Oranges	b/w	23 f.		15s.	B.I.F.
Life among the Bantu	b/w	39 f.		15s.	Foundation
Gold mining	b/w	40 f.		15s.	Foundation
South African goldfields		30 f.		42s.	C.G.
Riches of the Veld	b/w	46 f.		18s.	Rank
South Africa (2 parts)	b/w	46 f.	each	20s.	Hulton
Wine making	b/w	23 f.		12s.	B.I.F.
Diamond mining	b/w	47 f.		23s.	C.G.
Gold mining	b/w	47 f.		23s.	C.G.
Bushmen	b/w	27 f.		20s.	Hulton

Slides

Cape Town, Cape Peninsula, Port Elizabeth, Natal, Durban (5 sets)	4 sl.	per set	6s. 6d.	Sime

continued

Republic of South Africa	25 sl.	per set	85s.	Grant
The Cape	11 sl.			I.V.A.C.
Transvaal	22 sl.			I.V.A.C.
Transkei	8 sl.			I.V.A.C.
Kruger Park	6 sl.			I.V.A.C.
South West Africa	8 sl.			I.V.A.C.

Wallmaps and charts

South Africa (6 charts)	37 x 25 in.	per set	16s. 6d.	E.P.
Orange growing	25 x 37 in.		2s. 6d.	E.P.
Union of South Africa	20 x 30 in.		free loan	C.E.W.C.
South Africa (43 photographs)			free loan	C.E.W.C.
South African industries (12 mounts)			free loan	C.E.W.C.
Black washable 2-sided (South Africa/Africa)	54 x 42 in.		110s.	Grant
Central and Southern Africa (Westermann)	85 x 70 in.		185s.	Grant
Physical	70 x 48 in.		110s.	Philip
Outline	7 x 5 in.	per 25	3s. 9d.	E. J. Arnold
Political	43 x 33 in.		42s.	Philip
Physical	43 x 33 in.		42s.	Philip

SOUTH AMERICA

Trade and Travel Publications, 19 Leadenhall St., 6th Floor, London E.C.3
Producers of *South American Handbook,* an up-to-date guide. Approximately 25s.

Films

Geography of S. America	24 min.	*h.*	50s.	Gateway

Filmstrips

South America	b/w	45 f.	20s.	Hulton
General survey	b/w	40 f.	15s.	V.I.S.

Booklet

Timbers of South America	8s.	T.R.A.D.A.

Wallmaps and charts

Political	47 x 70 in.		110s.	Philip
Physical	47 x 70 in.		110s.	Philip
Outline map	7 x 5 in.	per 25	3s. 9d.	E. J. Arnold
Westermann map	54 x 74 in.		175s.	Grant
Black outline	30 x 40 in.		12s. 6d.	Hulton
Wall Atlas (5 maps)—Relief, climate, vegetation, political, population	43 x 33 in.	per set	170s.	
		per map	42s.	Philip
Graphic relief	34 x 42 in.		42s.	Philip

SOUTHERN YEMEN (SOUTH ARABIA)

Embassy of the Peoples' Republic of Southern Yemen, 95 Park Lane, London W.1
Free information service.

Film

Aden	b/w	12 min.	*h.*	15s.	Gateway

Filmstrip

Aden and the Suez Canal	44 f.	30s.	Rank

Slides

Aden (set of 4)	per set	6s. 6d.	Sime
Aden—general views (1963)		free loan	C.I.

Wallchart

Aden and Arabian Peninsula	20 x 30 in.	2s. 6d.	E.P.

SPAIN

Spanish Embassy, 24 Belgrave Square, London S.W.1
Limited literature in English (mainly tourist brochures).
Spanish National Tourist Office, 70 Jermyn St., London S.W.1
Free leaflets and literature of a tourist nature.
Barmerlea Book Sales Ltd., Annandale, North End Rd., London N.W.11
Suppliers of *Atlas Bachillerato, Universal y España.*
Barclays Bank
Economic report (ask for whole set).
Instituto Nacional de Estadistria, Ferraz 41, Madrid
Catalogue of statistical publications.

Maps—topographic

Selection 1:50,000 series:

San Sebastian 64	Tolosa 85
Pontevedra 185	Torquemada 274
Baltanas 312	Segovia 483
Almeria 1045	Malaga 1053

Very challenging and interesting cartography. About 10s. each from Stanfords.
Also 1:100,000 and 1:200,000 series.

Atlases

Nuevo Atlas de España	70s.
Atlas Commercial de España (1963)	240s.

The above are available from Stanfords.

Films

People of Spain	b/w	15 min.	*h.*	20s.	Rank
Spanish Meseta		9 min.	*h.*	25s.	Gateway
Spain and Portugal		16 min.	*h.*	50s.	Gateway
Village in Spain		20 min.	*h.*	40s.	Foundation

Filmstrips

N.W. Spain and east coast		34 f.		30s.	Rank
Central and south Spain		38 f.		30s.	Rank
Everyday life in Spain				36s.	E.P.
Spain 1				36s.	E.P.
Spain 2				36s.	E.P.
History, climate, physical	b/w			15s.	V.I.S.
People and economy	b/w			15s.	V.I.S.
Survey of Spain	b/w	44 f.		12s. 6d.	Foundation
Spain	b/w	40 f.		18s.	Unicorn
Introduction to Spain		25 f.		42s.	C.G.
Life in N.E. Spain		37 f.		30s.	Foundation
Spain and Portugal	b/w	43 f.		26s.	C.G.
Spain	b/w			20s.	Hulton

Slides

Madrid, Barcelona, etc. (a wide selection)	8 sl.	per set	15s. 6d.	Sime
Barcelona, Granada, Madrid (3 sets)	9 sl.	per set	13s. 6d.	Star
Set of geographical slides	50 sl.	per set	260s.	Harrap
Spain	35 sl.			I.V.A.C.

Wallmap

Iberian Peninsula (Westermann—in German)	68 x 67 in.	175s.	Grant

SUDAN

Maps—topographic

List of available maps from Stanfords.

Filmstrips

The Sudan	b/w		21s.	E.P.
A village in S. Sudan	b/w	29 f.	20s.	Hulton
Cotton from the Sudan	b/w	47 f.	20s.	Hulton
The wealth of Sudan		22 f.	25s.	Hulton

SWEDEN

The Swedish Institute for Cultural Relations, 49 Egerton Crescent, London S.W.3

Booklets and leaflets on Sweden available on request from teachers.

Films from S.S.

Gumperts Skolmateriel, P.O. Box 346, Göteborg 1

Very useful catalogue.

Recommended filmstrips (produced in Scandinavia):

1. Norway 2. Sweden 3. Denmark 4. Finland.

Kartforlaget GLA, Vasagatan 16, Stockholm 1
Massive catalogue of regional maps on many scales for Sweden and Scandinavia. Catalogue has many map examples.
Skolbokcentralen, Box 81, Johanneshov 1, Stockholm
Suppliers of atlases, including:
Nordisk Skolatlas—Elementary edition 10s.; Secondary edition 20s.
Garland Statisk Atlas 15s.
Nelson Rosen Skolatlas 10s.
Roths Atlas 10s.
Swedish National Map and Printing Organisation,
(Svenska Reproduktions Aktielbolget), Hudiksvallsgatan 4, Stockholm 6.
This is the address for enquiries; maps are available from Stanfords.
Recommended 1:50,000 sheets:
Skane: Malmo C2 NW; Halsingborg C3 SW; Tomelilla D2 NE; Kristianstad D3 SW.
Central Lakes Area: Hjo E8 SW: Linkoping F8 SE.
Stockholm: I10 NE; I10 SE; I10 NW; I10 SW.
Middle North: Kramfors I18 NW.
North: Gallivare K28.

Films

Sweden	b/w	18 min.	*h.*	20s.	Rank
Rhythm of a city	b/w	16 min.	*h.*	15s.	Foundation
Green gold (about forests)	b/w	23 min.		free loan	T.R.A.D.A.
Newspaper run		20 min.		free loan	P.F.B.

Filmstrips

Sweden			36s.	E.P.
Country and economy	b/w		15s.	V.I.S.
Sweden			30s.	Hulton
Sweden		31 f.	30s.	Rank
Sweden	b/w	43 f.	20s.	Foundation
Timber in Sweden	b/w	29 f.	18s.	Unicorn
Ports of industrial Scandinavia	b/w	42 f.	18s.	Rank
Rural life in Central Sweden		26 f.	42s.	C.G.

Slides

Millesgarden, Varmland, Stockholm	10 sl.	per set	17s. 6d.	Star
Sweden	30 sl.			I.V.A.C.

SWITZERLAND

Swiss National Tourist Office, Swiss Centre, 1 New Coventry St., London W.1
Travel and geographical information. Film catalogue available. Very large selection. Excellent.

Kummerley and Frey, Hallerstrasse 6-10, Bern
Atlas: *Schweizerischer Schulatlas,* about 40s.
Maps, books etc.; service very good.
Ask for catalogue.
Helbing und Lichtenhalm, Buchhandlung und Verlag, Friestrasse 40, Basel
Schoolbooks—catalogue available.
Recommended: *Schweizer Geographie* by Paul Vosseler, about 15s.

Maps—topographic

1:50,000 series. Good selection:

Tarasp 249	Rochers de Naye 262	Martigny 282
Interlaken 254	Arolla 283	

List of available maps from Stanfords.

Films

Switzerland	9 min.	*h.*	25s.	Rank
Village of Switzerland	16 min.	*h.*	30s.	Foundation

Filmstrips

Everyday life in Alpine Switzerland			36s.	E.P.
Mountain farmers	b/w	54 f.	18s.	Rank
Physical	b/w		15s.	V.I.S.
People and work	b/w		15s.	V.I.S.
Home life in Switzerland		60 f.	55s. 3d.	McGraw
Switzerland and the world		33 f.	15s.	Unicorn
Switzerland	b/w	40 f.	18s.	Unicorn
Physical	b/w	36 f.	20s.	Hulton
Transport	b/w	29 f.	20s.	Hulton
Farming	b/w	38 f.	20s.	Hulton
Alpine family	b/w	34 f.	20s.	Hulton
Amazing railways	b/w	43 f.	20s.	Hulton
History	b/w	36 f.	20s.	Hulton
Switzerland	b/w	44 f.	25s.	C.G.
Alpine village		27 f.	44s.	C.G.
Central Switzerland		23 f.	30s.	Hulton

Slides

Switzerland	8 sl.	per set	14s. 6d.	E.P.
Switzerland, Bernese Oberland, Tessin, Zurich, Bern, Lucerne, Jungfrau (7 sets)	10 sl.	per set	17s. 6d.	Sime

Also from Sime: very large selection of slides made on the continent.

Geographic regions	24 sl.	I.V.A.C.
Transport, industries and dwellings	20 sl.	I.V.A.C.

Wallmaps

Black washable 2-sided (Switzerland/Apennines)	54 x 42 in.	110s.	Grant

SYRIA

Filmstrip

Everyday life in Syria and Lebanon		36s.	E.P.

Slides

Syria and Jordan	25 sl.		I.V.A.C.

TANZANIA

Maps—topographic, etc.

1:250,000 Topographic series
1:2,000,000 Physical
1:25,000 Dar-es-Salaam and district
1:2,500 Dar-es-Salaam—Central area (4 sheets)
All available from Stanfords.

Filmstrips

Men against the bush	b/w		20s.	Hulton
Agriculture; cotton; coffee; sisal; tea; livestock; geographical background (7 strips)			free loan	C.I.

Slides

Tanzania—general views (1967)	free loan	C.I.
Dar-es-Salaam (1962)	free loan	C.I.

Wallmaps and charts

Physical and communications	47 x 36 in.	44s.	Philip
Official survey map	15 x 14 in.	1s. 3d.	C.I.
Relief and communications	47 x 37 in.	12s.	C.I.
Sisal plantation		8d.	C.I.

THAILAND

Barclays Bank
Economic report (ask for whole set)

Maps—topographic, etc.

1:250,000 Topographic series (32 sheets)
1:1,200,000 Physical (6 sheets)
1:10,000 Bangkok (8 sheets)
Available from Stanfords.

Films

Land of rice		14 min.	*h.* 40s.	Rank
Thailand	b/w	14 min.	*h.* 20s.	Foundation

Filmstrips

Life in Thailand	40 f.	55s. 3d.	McGraw
Plain of Thailand	29 f.	30s.	Rank
Everyday life in Thailand	35 f.	36s.	E.P.

Slides

Bangkok	4 sl.	per set	6s. 6d.	Sime
Thailand	32 sl.			I.V.A.C.

TIBET

Film

Tibetan traders	22 min.	*h.* 40s.	Foundation

TURKEY

Barclays Bank
Economic report (ask for whole set)

Maps–topographic, etc.

1:800,000 Topographic series }
1:3,000,000 Climate } when available
1:2,000,000 Earthquakes }
1:800,000 Economic (3 sheets)
1:10,000 Ankara Plan
Available from Stanfords.

Films

Turkey–emergent nation	11 min.	*h.* 25s.	Rank
Turkey–the bridge (1966)	19 min.	free loan	P.F.B.

Filmstrips

Gulem of Turkey		40 f.	55s. 3d.	McGraw
Turkey	b/w	34 f.	20s.	Hulton
Village and city life			30s.	Britannica

Slides

Istanbul, Ankara (2 sets)	4 sl.	per set	6s. 6d.	Sime

UGANDA

Uganda High Commission, Uganda House, Trafalgar Square, London W.C.2
Information in answer to questions.
Church Missionary Society
Exhibitions, sets of information for discussion, filmstrips, slides and topic box.

Maps–topographic, etc.

1:1,000,000 General
1:500,000 Physical
1:250,000 Topographic series
1:10,000 Township maps: Entebbe, Jinja, Kampala, Mbale
1:250,000 National parks: Queen Elizabeth; Murchison Falls
1:250,000 Soil series
1:250,000 Vegetation series
1:250,000 Geology series
All about 14s. each. Available from Stanfords.

Slides

Cotton and coffee (1964)	free loan	C.I.
People of Uganda (1964)	free loan	C.I.
Scenery (1964)	free loan	C.I.
Towns (1964)	free loan	C.I.

Wallmaps and charts

Physical & communications	36 x 46 in.	44s.	Philip
Uganda becomes a nation	30 x 40 in.	4s.	H.M.S.O.
Charts (6 panels)	12 x 15 in.	free loan	C.E.W.C.
Official survey map	15 x 14 in.	1s. 3d.	C.I.
Photographic set (12 views)	12 x 15 in.	11s. 6d.	C.I.
Relief and communications	37 x 44 in.	12s.	C.I.

UNITED ARAB REPUBLIC

see Egypt

UNITED NATIONS

see also under individual countries.

U.K. Committee for UNICEF, New Gallery Centre, 123 Regent St., London W.1

Books: *UNICEF and Britain; Questions and answers on the needs of children and UNICEF.*
Slide sets for loan (postage payable). List.
Filmstrips for loan (postage payable) or purchase (20s. for 38 f.). List.
Facts and figures sheet. List of display materials for hire at very reasonable rates.

†United Nations Information Centre, 14-15 Stratford Place, London W.1

A broad spectrum of information readily and cheaply available (no material on continents or individual countries).
Films and filmstrips (catalogue). Pamphlets explaining U.N., F.A.O., U.N.H.C. Refugees, U.N.W.R.A., W.H.O.
Reference library (loan service). Photographic library.

Council for Education in World Citizenship

Poster sets (9) on United Nations. (Service only available to member schools.)

UNITED STATES–GENERAL

United States Information Service, Grosvenor Square, London W.1

Assistance to students and teachers in Grammar Schools and Universities.

United States Travel Service, 22 Sackville St., London W.1

National Coal Association, Coal Building, 1130 Seventeenth Street NW, Washington D.C. 20036

Useful source of pictures and comics designed for U.S. schools.

American Forest Products Ind. Inc., 1885 K. Street NW, Washington 20006

Large range of booklets and maps, including *The story of hardboard, The story of plywood, It's a tree country* (summary of U.S. forests).
Project sheet and maps on tree and forest growth and use.

A. J. Nystrom & Co., 3333 Elston Avenue, Chicago, Illinois 60618
Suppliers of relief maps and transparencies.
Materials for use with overhead projector.
Catalogue available. Good for U.S.A. topics.

H. E. Budek Company Inc., 324 Union St., Hackensack, New Jersey
American colour slide producers.
Catalogue of slides of North America available.

Petroleum Information Bureau
Pamphlet: *Oil–North America.*

International Harvester Export Company, 401 North Michigan Avenue, Chicago, Illinois 60611
General products catalogue and information on farming equipment—colour and photographs really excellent.

†C. F. & I. Steel Company, P.O. Box 316, Pueblo, Colorado 81002
Folder of information on U.S. Steel Industry.
(Single copies.)

Maps—topographic

A very wide range of maps available. Consult Stanfords on your needs.

Atlases

Oxford Regional Economic Atlas of U.S. and Canada			75s.	O.U.P.

Film cassettes (8 mm)

Irrigation farming				
Sheep ranching				
Logging			each 69s.	Gateway

Filmstrips

United States Community	b/w	46 f.	18s.	Rank
Everyday life in the U.S.A.: Farming; Industry			each 36s.	E.P.
Farm family life in the U.S.A.	b/w	35 f.	29s. 9d.	McGraw
Cotton growing		27 f.	15s.	Textile
North America	b/w	45 f.	20s.	Hulton
Beef cattle in the U.S.A.		30 f.	42s.	C.G.
Oxford Projected Atlas—The Americas			34s.	C.G.
Miners of the Americas	b/w		23s.	C.G.
Natives of the Americas	b/w		23s.	C.G.

Slides

Slide-o-map set on N. America		24 sl.		
Slide-o-map set on U.S.A. and Canada		25 sl.		

continued

Countryside I	18 sl.		
Countryside II	22 sl.		
Towns and transport I	26 sl.		
Towns and transport II	49 sl.		All from I.V.A.C.

Very large selection from Sime (catalogue).

Wallmaps and charts

Black washable 2-sided (N. America/S. America)	54 x 42 in.	110s.	Grant
U.S.A. & S. Canada–Physical (Westermann)	81 x 80 in.	185s.	Grant
North America–Political	48 x 59 in.	110s.	Philip
North America–Physical	47 x 59 in.	110s.	Philip
Physical/Political (made in U.S.A.)	64 x 80 in.	180s.	Philip
Face of America (84 mounts)	15 x 18 in.	free loan	C.E.W.C.
Agriculture, forestry and fisheries	38 x 50 in.	40s.	Grant
Mining, industry and transport	38 x 50 in.	40s.	Grant
Black outline	40 x 30 in.	12s. 6d.	Hulton
U.S. Budget; U.S. Government; Electing a president (poster set)		free loan	C.E.W.C.
North America–Physical	43 x 33 in.	42s.	Philip
North America–Political	43 x 33 in.	42s.	Philip
Graphic relief	42 x 34 in.	42s.	Philip

UNITED STATES–EAST

Niagara Falls Area, Chamber of Commerce, Niagara Falls, N.Y. 14303
Useful tourist literature about the Falls area.

Port of New York Authority, Public Affairs Department, 11 Eighth Avenue, 15th Street, N.Y. 10011
Very good, but please use only in special cases, specifying need.

Vermont Development Department, Montpellier, Vermont 05602
Maps and brochures of their area.

New Hampshire State Planning and Development, Concord, New Hampshire
Information sheets. List of books on New Hampshire.
State map. Farm facts.

Films

New York (BBC TV)	b/w	22 min.	*h.* 20s.	Foundation
Industrial Heart (BBC TV)	b/w	22 min.	*h.* 20s.	Foundation

Filmstrips

North East		36s.	E.P.
South East		36s.	E.P.
New England	50 f.	57s. 6d.	McGraw
Middle Atlantic States	50 f.	57s. 6d.	McGraw
The Great Lakes	37 f.	63s. 9d.	McGraw

Megalopolis		37 f.		63s. 9d.	McGraw
New England				42s.	C.G.
The Appalachians				30s.	Hulton
New England				30s.	Hulton
Maine harbour town	b/w	48 f.		18s.	Rank
Buffalo: Industrial Lake port	b/w	46 f.		18s.	Rank
Living in a metropolis	b/w	43 f.		18s.	Rank
New York		27 f.		42s.	C.G.
Industrial East	b/w	41 f.		23s.	C.G.

Slides

Washington, New York, North Carolina, Pennsylvania, Philadelphia (5 sets)		4 sl.	per set	6s. 6d.	Sime
The Eastern States		25 sl.	per set	85s.	Grant

Wallmap

New York (Westermann)	44 x 56 in.	120s.	Grant

UNITED STATES—CENTRAL AND SOUTHERN

State of Missouri, Division of Commerce and Industrial Development, 8th Floor, Jefferson Building, Jefferson City, Missouri 65102

Tourist literature and maps. Photographs. State maps. Fact sheets.

†Tennessee Valley Authority, Knoxville, Tennessee

Very good study set.

Oklahoma Planning and Resources Board, Oklahoma City, Chamber of Commerce, Oklahoma

Pamphlets on history; urban development (1889-1989); Indians; cultural activities.

City Guide A-Z. State Highway Map. Book of colour pictures.

Texas Highways Department, Austin, Texas. Box 5064

Vacation literature. Official State Highway Map. Texas facts.

Films

Cattle and the corn belt	b/w	20 min.	*h.* 20s.	Rank
Changing cotton lands	b/w	20 min.	*h.* 20s.	Rank
The Mid West	b/w	22 min.	*h.* 20s.	Foundation

Filmstrips

Mountain states		36s.	E.P.
Central lowlands		36s.	E.P.
Middle West	50 f.	57s. 6d.	McGraw
Central farming region	37 f.	63s. 9d.	McGraw
The Great Plains	37 f.	63s. 9d.	McGraw
The Gulf Coast	37 f.	63s. 9d.	McGraw
Middle West		30s.	Hulton
Southern Interior		30s.	Hulton

continued

Changing cotton lands	b/w	40 f.		18s.	Rank
The cotton lands	b/w	43 f.		23s.	C.G.
Gulf Coast				30s.	Hulton
The Mississippi	b/w	30 f.		15s.	Foundation
The Mississippi Basin	b/w	37 f.		25s.	C.G.
Florida				30s.	Hulton
Southern United States		50 f.		57s. 6d.	McGraw

UNITED STATES–WESTERN

Films

Through the South West	b/w	22 min.	*h.*	20s.	Foundation
California (BBC TV)	b/w	22 min.	*h.*	20s.	Foundation
Rocky mountains (BBC TV)	b/w	22 min.	*h.*	20s.	Foundation
North Alaska		19 min.	*h.*	20s.	P.F.B.
Life in a Mediterranean climate (California)	b/w	10 min.	*h.*	15s.	Gateway

Filmstrips

Pacific coastlands				36s.	E.P.
South West United States		50 f.		57s. 6d.	McGraw
California fruit ranch				42s.	C.G.
Lumber States	b/w	49 f.		18s.	Rank
Mountain States				30s.	Hulton
Pacific coast		50 f.		57s. 6d.	McGraw
California				30s.	Hulton
California				44s.	C.G.

Slides

Arizona, Grand Canyon, Alaska, California, San Francisco (5 sets)	4 sl.	per set	6s. 6d.	Sime
The Western States	25 sl.	per set	85s.	Grant
San Francisco, Yosemite, Grand Canyon (3 sets)	5 sl.	per set	7s. 6d.	Star

Wallchart

Peach growing in California	25 x 37 in.	2s. 6d.	E.P.

U.S.S.R.

Society for Cultural Relations with the U.S.S.R., 118 Tottenham Court Rd., London W.1

Reference library; visual aids department. Very long list of b/w filmstrips; also a series of special geography filmstrips and maps. Six maps (40 x 25 in.) including: U.S.S.R. 1917-67; Large power plant; Schematic map; Agriculture; Economic development; plus 79 pages of information, 16s. 6d. Large library of photographs available for loan. Loans are for up to one month. Postage payable.

Booklet

Oil in the U.S.S.R.	free	P.I.B.

Atlases

Soviet Union in maps	6s.	Philip
Oxford Regional Economic Atlas of U.S.S.R. & East Europe	63s.	O.U.P.

Films

Dateline Russia		18 min.	free loan	Ford
Soviet Union—land and people	b/w	17 min.	*h.* 25s.	Gateway
Russia—an introduction		21 min.	*h.* 50s.	Gateway
Collective farm in Central Asia	b/w	20 min.	*h.* 25s.	McGraw
New town in Siberia	b/w	20 min.	*h.* 25s.	McGraw
Landscape and climate	b/w	20 min.	*h.* 25s.	McGraw
Russia in Europe		19 min.	*h.* 30s.	Foundation

Long list for hire from E.T.V., including geography, travel, science, industry, agriculture and cultural topics.

Filmstrips

Geographical survey I	b/w		21s.	E.P.
Geographical survey II	b/w		21s.	E.P.
Russia—Physical	b/w		15s.	V.I.S.
Russia—Human	b/w		15s.	V.I.S.
Soviet Russia (1965-66)	b/w	58 f.	18s.	Unicorn
Soviet Russia	b/w	47 f.	20s.	Hulton
The Ukraine	b/w	34 f.	23s.	C.G.

Long list of Russian regional filmstrips from Collet's Russian Shop, 39 Museum St., London W.C.1, including:

Black Sea coast	European R.S.F.S.R.	
East Siberia	Leningrad	
Natural zones	Soviet Georgia	each 28s. 6d.

Slides

North, Central and South (3 sets)	12 sl.	per set	41s.	Grant
Moscow	10 sl.	per set	15s.	Sime
Leningrad, Kiev, Samarkand, Yalta (4 sets)	4 sl.	per set	6s. 6d.	Sime
Moscow	9 sl.	per set	13s. 6d.	Star
Moscow	13 sl.			I.V.A.C.
North West Russia	14 sl.			I.V.A.C.
Leningrad	13 sl.			I.V.A.C.
Rural life (Ukraine)	15 sl.			I.V.A.C.
Georgia and Azerbaijan	12 sl.			I.V.A.C.

Wallmaps and charts

Soviet industries	30 x 40 in.	8s. 6d.	Gateway

continued

North Asia/U.S.S.R. (Westermann)		96 x 76 in.	185s.	Grant
Moscow (Westermann–in German)		44 x 56 in.	120s.	Grant
Physical/Political		64 x 44 in.	130s.	Philip

VENEZUELA

Timber Research and Development Association

Notes on Venezuelan timbers–free.

Maps–topographic, etc.

1:2,000,000 Political
1:1,000,000 Physical/Political (8 sheets)
1:1,000,000 Geological
All from Stanfords.

Filmstrips

Venezuela		31 f.		24s.	Unicorn
Venezuela	b/w			15s.	V.I.S.
Venezuela		43 f.		57s. 6d.	McGraw

Slides

La Guaira, Caracas (2 sets)		4 sl.	per set	6s. 6d.	Sime
Venezuela looks ahead				free loan	P.F.B.

WEST INDIES AND CARIBBEAN

Commonwealth Institute

Large selection of maps and charts available. Slides on free loan.

Voluntary Committee on Overseas Aid and Development

List of maps, charts, films, etc.

Atlases

Caribbean Modern School Atlas		16s.	Philip
Supplements only	per ten	30s.	Philip
Visual Atlas–West Indies Edition		8s.	Philip

Maps–topographic

Produced by the Directorate of Overseas Surveys for former British islands; maps also available for French, American and Dutch islands. All from Stanfords.

Booklet

Oil–Latin America and the Caribbean P.I.B.

Film

Sponges	19 min.	free loan	P.F.B.

Filmstrips

The British Caribbean		36s.	E.P.
The West Indies: Climate, vegetation, peoples	b/w	15s.	V.I.S.

continued

Products	b/w			15s.	V.I.S.
Jamaica	b/w	29 f.		20s.	Hulton
The West Indies Region	b/w	41 f.		25s.	C.G.
Jamaica and the sugar industry				44s.	C.G.

Slides

Sets (4 slides each) including Trinidad; Barbados; Nassau; Grenada; Jamaica; Havana; Kingston		per set	6s. 6d.	Sime

Wallmaps and charts

The Caribbean	30 x 40 in.		free loan	C.E.W.C.
British Honduras	30 x 40 in.		2s.	H.M.S.O.
Trinidad becomes a nation	30 x 40 in.		2s.	H.M.S.O.
Jamaica	30 x 40 in.		2s.	H.M.S.O.
The West Indies	16 plates		15s.	H.M.S.O.
The Commonwealth in the Caribbean (4 charts)	25 x 37 in.	per set	10s.	E.P.
Banana growing in Jamaica	25 x 37 in.		2s. 6d.	E.P.
West Indies & Caribbean—Political	48 x 37 in.		44s.	Philip
West Indies—Physical & communications	46 x 36 in.		44s.	Philip
Jamaica—Physical & communications	40 x 31 in.		44s.	Philip

YUGOSLAVIA

Yugoslav National Tourist Office, 143 Regent St., London W.1
Barclays Bank
Economic report (ask for whole set).

Filmstrips

Across the Yugoslav karst		44s.	C.G.
Yugoslavia		30s.	Hulton
Yugoslavia	b/w	20s.	Hulton

Slides

Dalmatian Coast	10 sl.		15s.	Star
Koper, Split, Dubrovnik, Trogir, Ljubljana, Bled, Opatija, Rijeka, Novi Vinodolski, Mostar, Plitvice, Beograd, Cetinje, Zadar (14 sets)	4 sl.	per set	6s. 6d.	Sime

ZAMBIA

High Commissioner for the Republic of Zambia, 7-11 Cavendish Place, London W.1
Monthly booklet *Zambia* contains articles on development of many aspects of Zambia.

National Westminster Bank
Economic report

Maps–topographic, etc.

1:1,000,000 Topographic series (5 sheets)
1:250,000 Topographic series
1:20,000 Lusaka town plan
1:2,500,000 Minerals map
Available from Stanfords.

Filmstrips

African life in copper belt	b/w			free loan	C.I.
Copper mining in Zambia	b/w			free loan	C.I.
Agricultural improvements				free loan	C.I.

Slides

Kruger National Park	4 sl.	per set	6s. 6d.	Sime
Story of copper: Prospecting; production; mines			free loan	C.I.
General slides (1966)			free loan	C.I.

Map

Official survey map	15 x 14 in.	1s. 3d.	C.I.

NAMES AND ADDRESSES ABBREVIATED IN THE TEXT

A.E.I.	Associated Electrical Industries Ltd. Film Library, 33 Grosvenor Place, London S.W.1.
E. J. Arnold	Arnold, E. J. and Son Ltd., Butterley Street, Leeds 10.
	Association of Agriculture, 78 Buckingham Gate, London S.W.1.
Barclays	Barclays Bank Ltd., Economic Intelligence Division, 54 Lombard Street, London E.C.3.
Boulton	Boulton Hawker Films, Hadleigh, Suffolk.
B. & F.	British and Foreign Bible Society, 146 Queen Victoria Street, London E.C.4.
B.I.F.	British Instructional Films Ltd., 2 Dean Street, London W.1.
B.S.C.	British Steel Corporation Film Library, Wilton Crescent, Merton Park, London S.W.19.
B.T.	British Transport Films, Melbury House, Melbury Terrace, London N.W.1.
Britannica	Foundation Film Library, Brooklands House, Weybridge, Surrey.
B. Wool	British Wool Marketing Board, Education Department, Oak Mills, Clayton, Bradford, Yorkshire.
C.O.I.	Central Office of Information, Hercules Road, London S.E.1.
C.O.I. (C.F.L.)	Central Office of Information (Central Film Library), Government Buildings, Bromyard Avenue, London W.3.
Cheshire	Cheshire Films, Rendell and Wilson, 176a Ashley Road, Hale, Altrincham, Cheshire.
C.M.S.	Church Missionary Society, 6 Salisbury Square, London E.C.4.
C.G.	Common Ground Ltd., 44 Fulham Road, London S.W.3.
C.I.	Commonwealth Institute, Kensington High Street, London W.8.
Cochrane	Cochrane, John, Pty. Ltd., 373 Bay Street, Port Melbourne, Victoria, Australia, 3207.
Contemporary	Contemporary Films Ltd., 14 Soho Square, London W.1.
C.E.W.C.	Council for Education in World Citizenship, 93 Albert Embankment, London S.E.1. (*Service only available to member schools.*)
Dunlop	Dunlop Film Library, Wilton Crescent, Merton Park, London S.W.19.

E.F.V.A.	Educational Foundation for Visual Aids, 33 Queen Anne Street, London W.1.
E.P.	Educational Productions Ltd., East Ardsley, Wakefield, Yorkshire.
E.S.A.	Educational Supply Association, Pinnacles, Harlow, Essex.
Esso	Esso Petroleum Co. Ltd., Education Service, Victoria Street, London S.W.1.
E.T.V.	Educational and Television Films Ltd., 2 Doughty Street, London W.C.1.
European	European Schoolbooks Ltd., 100 Great Russell Street, London W.C.1.
Ford	Ford Film Library, Wilton Crescent, Merton Park, London S.W.19.
Foundation	Foundation Film Library, Brooklands House, Weybridge, Surrey.
Gateway	Gateway Educational Films Ltd., 470 Green Lanes, Palmers Green, London N.13.
Geographia	Geographia Ltd., 178 Great Portland Street, London W.1.
Grant	Grant Educational Co. Ltd., 91-95 Union Street, Glasgow C.1.
Gunasena	Gunasena Ltd., Colombo, Ceylon.
Harrap	Harrap, George G. and Co. Ltd., P.O. Box 70, 182 High Holborn, London W.C.1.
H.M.S.O.	Her Majesty's Stationery Office, P.O. Box 569, London S.E.1.
Herring	Herring Industry Board, 1 Glenfinlas Street, Edinburgh 3.
Hulton	Hulton Educational Productions, Alan House, 55-59 Saffron Hill, London E.C.1.
India	India Films, India House, Aldwych, London W.C.2.
I.V.A.C.	International Visual Aids Centre, 691 Chaussée de Mons, Brussels 7.
I.W.S.	International Wool Secretariat, Department of Education, Wool House, Carlton Gardens, London S.W.1.
J.K.F.	Jomo Kenyatta Foundation, Nairobi, Kenya.
Longman	Longman Group Ltd., Longman House, Burnt Mill, Harlow, Essex.
Macmillan	Macmillan and Co. Ltd., 4 Little Essex Street, London W.C.2.
McGraw	McGraw-Hill Publishing Co. Ltd., McGraw-Hill House, Shoppenhangers Road, Maidenhead, Berks.
M.M.S.	Methodist Missionary Society, 25 Marylebone Road, London N.W.1.
Midland	Midland Bank Ltd., Poultry, London E.C.2.

Min. of Ag.	Ministry of Agriculture, Fisheries and Food, The Library, Whitehall Place, London S.W.1.
Murray	Murray, John, (Publishers) Ltd., 50 Albemarle Street, London W.1.
National	National Westminster Bank Ltd., P.O. Box 14, 1 Princes Street, London E.C.2.
Nelson	Nelson, Thomas, and Sons Ltd., 36 Park Street, London W.1.
N.Z.H.C.	New Zealand High Commissioner, New Zealand House, Haymarket, London S.W.1.
Oxfam	Oxfam, 274 Banbury Road, Oxford.
O.U.P.	Oxford University Press, Ely House, 37 Dover Street, London W.1.
Pergamon	Pergamon Press Ltd., Headington Hill Hall, Oxford OX3 OBR.
P.F.B.	Petroleum Film Bureau, 4 Brook Street, London W.1.
P.I.B.	Petroleum Information Bureau, 4 Brook Street, London W.1.
Philip	Philip, George, and Son Ltd., 12-14 Long Acre, London W.C.2.
P. & T.	Philip & Tacey Ltd., 69 Fulham High Street, London S.W.6.
Pictorial	Pictorial Charts Educational Trust, 132-138 Uxbridge Road, West Ealing, London W.13.
Potato	Potato Marketing Board, 50 Hans Crescent, London S.W.1.
Rank	Rank Film Library, Great West Road, Brentford, Middlesex.
Sime	Sime, David, 11 Balmoral Drive, Bearsden, Glasgow.
S.C.	Slide Centre Ltd., Portman House, 17 Brodrick Road, London S.W.17.
S.S.	Sound Services Ltd., Wilton Crescent, Merton Park, London S.W.19.
Stanfords	Stanford, Edward, Ltd., 12-14 Long Acre, London W.C.2.
Star	Star Press Ltd., Venice Chambers, 61 Lord Street, Liverpool 2.
S. Times	Sunday Times, 200 Grays Inn Road, London W.C.1.
Textile	Textile Council, 3 Alberton Street, Manchester 3.
T.R.A.D.A.	Timber Research and Development Association, Hughenden Valley, High Wycombe, Bucks.
Unicorn	Unicorn Head Productions Ltd., 42 Westminster Palace Gardens, Artillery Row, London S.W.1.
Unilever	Unilever Educational Productions, Unilever House, London E.C.4.
U.K.C.F.H.C.	United Kingdom Committee for the Freedom from Hunger Campaign, 17 Northumberland Avenue, London W.C.2.
U.P.S.	United Publishers Services (Hong Kong) Ltd., Stanhope House, 7th Floor, 734 King's Road, Hong Kong.

V.I.S.	Visual Information Service, 12 Bridge Street, Hungerford, Berks.
V.C.O.A.D.	Voluntary Committee on Overseas Aid and Development, 69 Victoria Street, London S.W.1.
Walton	Walton Sound and Film Services Ltd., 87 Richford Street, London W.6.
Woodmansterne	Woodmansterne Ltd., Watford WD1 8RD.

PUO

Sale: 50p